QUEUEING THEORY
AND
APPLICATIONS

PROCEEDINGS OF THE ARAB SCHOOL OF SCIENCE AND TECHNOLOGY

J. Albaigés (editor)
Marine Pollution

R. Descout (editor)
**Applied Arabic Linguistics
and Information and Signal Processing**

A. H. El-Abiad (editor)
Power Systems Analysis and Planning

C. Foulard (editor)
**Product Development and Production Engineering
in Manufacturing Industries**

T. Kailath (editor)
Modern Signal Processing

P. MacKay (editor)
Computers and the Arabic Language

P. D. T. O'Connor (editor)
Reliability Engineering

S. Özekici (editor)
Queueing Theory and Applications

E. E. Pickett (editor)
A. Habal and F. Abosamra (co-editors)
Atmospheric Pollution

G. Warfield (editor)
Solar Electric Systems

FORTHCOMING

R. Risebrough (editor)
Pollution and the Protection of Water Quality

QUEUEING THEORY AND APPLICATIONS

Edited by

S. ÖZEKİCİ
Boğaziçi University, Turkey

⬤HEMISPHERE PUBLISHING CORPORATION
A member of the Taylor & Francis Group

New York   Washington   Philadelphia   London

QUEUEING THEORY AND APPLICATIONS

1 2 3 4 5 6 7 8 9 0   B R B R   9 8 7 6 5 4 3 2 1 0

Cover design by Sharon DePass.
A CIP catalog record for this book is available from the British Library.

Library of Congress Cataloging-in-Publication Data

Queueing theory and applications / edited by Süleyman Özekici.
      p.   cm.—(Proceedings of the Arab School of Science and Technology)
   Includes bibliographical references.

   1. Queueing theory. I. Özekici, Süleyman. II. Series: Proceedings of the Arab School on Science and Technology.
T57.9.Q479   1990
519.8'2—dc20   90-4241
                                                    CIP

ISBN 0-89116-995-4

CONTENTS

CONTRIBUTORS

Süleyman Özekici
Department of Industrial Engineering
Boğaziçi University
Bebek, Istanbul, Turkey

Walid Moussattat
Higher Institute of Applied Science
and Technology
Damascus, Syria

M. Akif Eyler
Department of Industrial Engineering
Bilkent University
Ankara, Turkey

Ali R. Kaylan
Department of Industrial Engineering
Boğaziçi University
Bebek, Istanbul, Turkey

Nihal Pekergin
Ecole des Hautes Etudes en Informatique
Universite René Descartes
75006, Paris, France

Mohammad S.Obaidat
Department of Electrical
and Computer Engineering
Univ. of Missouri-Columbia/Kansas City
Independence, MO 64050,USA

Stanley R. Pliska
Department of Information
and Decision Sciences
University of Illinois at Chicago
Chicago, Ill. 60680, USA

Nesim Erkip & Sedef Meral
Department of Industrial Engineering
Middle East Technical University
Ankara, Turkey

PREFACE

The 4th winter session of the Arab School of Science and Technology was held in Damascus, Syria during December 13-19, 1988. The session was organized as a six-day intensive seminar on *Queueing Theory and Applications*, and I had the pleasant opportunity to work for the Arab School as a scientific consultant in cooperation with the scientific committee. The main objective of the session was to educate the participants on basic topics in queueing theory, and orient them towards research on current issues in theory and applications. The chapters that make up this book were prepared with this objective in mind. The proceedings of the seminar was extended and edited to present our readers this graduate level textbook and reference book on *Queueing Theory and Applications*.

The book consists mainly of two parts: Chapters 1-5 concentrate more on basic theory, while Chapters 6-10 emphasize applications. The first part includes a brief review of stochastic processes used in queueing theory (Chapter 1), stochastic modeling of elementary and advanced queueing systems (Chapters 2 and 3), simulation of queueing systems (Chapter 4), and queueing networks (Chapter 5). In the second part, the application oriented topics consist of performance modeling and simulation in computer systems (Chapters 6 and 7), academic computer networks (Chapter 8), management and optimization of queueing systems (Chapter 9), and finally, queueing network models for flexible manufacturing systems (Chapter 10). All presentations are made with a view toward applications in operations research and management science related fields. I believe that this outline is suitable for a one semester or two quarter graduate level course.

On behalf of all of the participants and lecturers in the seminar I would like to thank the staff of the Arab School for a fine organization. I would also like to recall my fond memories of Mr. Adnan Rıfai, the General Secretary, whose sudden and unexpected death only a few days before the session left us all with deep sorrow.

The book is typeset by TEX using the Macintosh software TEXtures, and the originals were printed by a LaserWriter Plus. Many people helped me in the preperation of the final copy. I would like to acknowledge the research assistants at the Department of Industrial Engineering of Boğaziçi University for their time-consuming efforts in entering the manuscripts in my computer files, and for proofreading the initial drafts. In particular, I feel deeply indepted to Erol Nalçacı for the long hours he spent in front of a Macintosh II.

Istanbul, November 1989　　　　　　　　　　　　　　　　　　　*Süleyman Özekici*

THE ARAB SCHOOL OF SCIENCE AND TECHNOLOGY

History

The Arab School of Science and Technology, a pan-Arab, non-profit organization headquartered in Damascus, Syria, was founded in 1978 by the initiative of the Kuwait Institute for Scientific Research (KISR), the Scientific Studies and Research Center (SSRC), and the Supreme Council of Sciences (SCS) in Syria to provide a high level continuing education program to Arab scientists in fields that are judged crucial to the development of the Arab countries.

Since its establishement, the School has dealt with four major topics: Electronics, Energy, Environment, and Informatics. Additionally, two sessions were held in Applied Mathematics related fields. The School has plans to expand into other areas of specialization, such as the transfer, adaptation, and development of technology in the Arab World.

The School has attempted to create for each topic a regular forum of scientific exchange in chosen areas of the Arab World. It has succeeded in establishing one for Electronics in Syria, and another one for Energy in Kuwait. The forum for the Environment series and the Informatics series were originally planned to be established in one of the Maghreb countries (i.e., Tunisia, Algeria, or Morocco), but until plans are finalized, the sessions in these series will be held in Syria.

In addition to its original main sponsors (KISR and SSCR), the School has also been sponsored by a number of national (such as the Center Nationale de Coordination et de Plannification de la Recherche Scientific et Tecnique, Morocco), regional (such as the Arab League Educational, Cultural, and Scientific Organization) and international (such as UNESCO/ROSTAS, UNDP, UNIDO, UNEDBAS, and ALECSO) organizations.

Objectives

The Arab School, through its sessions, aims to fulfill the following objectives:

1. Familiarizing Arab scientists, engineers, and university professors with the latest advances in science and technology through intensive advanced post-graduate and highly specialized post-doctoral courses given by leading scientists.
2. Facilitating direct contact between Arab scientists to establish a propitious atmosphere for joint cooperation in the field of science and technology.
3. Encouraging Arab scientists working abroad to return to their countries by providing them with opportunities to contribute to the School's activities, and closely review the scientific resources of their countries.

4. Facilitating scientific cooperation among Arab and Muslim countries by direct contacts provided by the School.

5. Providing an overview of scientific activities in Arab countries by publishing session proceedings which cover scientific and technological developments in the Arab/Muslim World.

ARAB SCHOOL AUTHORIZED COMMITTEES

Arab School Coordinator :

Dr. A. H. Said
Arab School of Science and Technology
P.O.Box: 7028
Damascus, Syria

Scientific Committee :

Dr. W. Moussattat
Higher Instute of Applied Science and Technology
Damascus, Syria

Dr. S. Khiyami
Higher Instute of Applied Science and Technology
Damascus, Syria

Dr. M. Farah
Higher Instute of Applied Science and Technology
Damascus, Syria

Dr. M. B. Mounajed
University of Damascus
Damascus, Syria

Organization Supervisor :

Mrs. G. Akbik
Arab School of Science and Technology
P.O.Box: 7028
Damascus, Syria

PREVIOUS AND FORTHCOMING SESSIONS

Previous Sessions	Date		Location
Solid State Electronics	Summer	1978	Syria
Communications	Summer	1979	Syria
Minicomputers, Microprocessors,and Their Applications	Summer	1980	Syria
Power Systems:Analysis and Planning	Winter	1981	Kuwait
Control Systems: Theory and Applications	Summer	1981	Syria
Solar Electricity Systems	Winter	1982	Kuwait
Pollution and Protection of the Water Quality	Summer	1982	Syria
Modern Signal Processing	Summer	1983	Syria
Applied Arabic Linguistics and Signal and Information Processing	Fall	1983	Morocco
Technology Transfer and Adaptation in the Arab World	Fall	1983	Syria
Informatics and Applied Arabic Linguistics	Summer	1985	Syria
Atmospheric Pollution	Summer	1985	Syria
Mathematical Modeling and Applications	Spring	1986	Syria
Reliability	Summer	1986	Syria
Marine Pollution	Summer	1987	Syria
Product Development and Production Engineering	Summer	1988	Syria
Queueing Theory and Applications	Winter	1988	Syria
Information and Computer Networks in the Arab States	Summer	1989	Syria

Forthcoming Sessions	Date		Location
Application of Remote Sensing to Hydrology and Water Resources	March	1990	Syria
Planning in the Scientific Research and Engineering Development Sectors	July	1990	Syria
Microelectronics (conference)		1990	Syria
Data Bases and Their Arabization		1990	Syria
Computer Aided Design		1991	
Composite Materials, Their Technology, and Applications		1991	
Sensors		1992	

Chapter 1

STOCHASTIC PROCESSES IN QUEUEING THEORY

Süleyman Özekici
Boğaziçi University
Department of Industrial Engineering
Bebek, Istanbul, Turkey

1. INTRODUCTION

Queueing theory is one of the most diversified fields of applied sciences that is widely used in a variety of areas including operations research, management science, computer science, telecommunications, and production systems, to name only a few. This richness and abundance in applications is properly reflected in the stochastic models, tools, and techniques used in analyzing them. The importance of queueing theory lies not only in its applicability to real-life problem with its computationally tractable procedures, but in the elegance and completeness of the underlying stochastic models and mathematics as well.

This Chapter presents a quick review of some of the important stochastic processes used in queueing theory. Our purpose is to provide our reader a reference chapter that complements the topics pertaining to stochastic modeling in this book. This is done in a rather informal style by emphasizing only those results which bear importance to operations researchers and management scientists. Thus, the processes are introduced with a view towards applications, especially in queueing theory, and application oriented theories including ergodic, potential, and renewal theory are emphasized. Most of the material presented is pretty standard, and the reader is referred to any one of the good textbooks on stochastic processes for details and proofs. We recommend, for example, Çınlar (1975), Ross (1983), or Bhat (1984) to our reader with an application oriented mind.

Poisson processes are covered in Section 2, and the following two sections, 3 and 4, are devoted to Markov chains and Markov processes, respectively. Renewal processes and renewal theory are the topics of Section 5, and, finally Section 6 discusses Markov renewal processes and Markov renewal theory.

2. POISSON PROCESSES

2.1. Definitions and Basic Results

Poisson processes are no doubt the most widely used customer arrival processes in queueing theory. This is due not only to their computational tractability associated with various probabilistic quantities of interest, but more to the fact that they naturally represent customer arrival streams . This observation is justified by the following three properties satisfied by a Poisson process:

1. Customers arrive one by one, rather than in groups or batches,
2. Customers make decisions independent of each other regarding their arrival,
3. Customers make decisions independent of time regarding their arrival.

As a matter of fact, these three qualitative properties are put into quantitative forms to make up the following formal definition of a Poisson process $A = \{A_t;\ t \geq 0\}$, where A_t is the total number of customer arrivals until time t:

1. $t \to A_t$ is right-continuous, and increasing by jumps of size 1 only $(A_0 = 0)$,
2. A has independent increments, in other words

$$P\{A_{t+s} - A_t = k | A_u;\ u \leq t\} = P\{A_{t+s} - A_t = k\}\ ,$$

3. A has stationary increments, in other words

$$P\{A_{t+s} - A_t = k\} = P\{A_s = k\}\ .$$

Moreover, it is well-known that these properties lead to the Poisson distribution

$$P\{A_s = k\} = \frac{e^{-\lambda s}(\lambda s)^k}{k!}\ , \qquad k = 0, 1, 2, \ldots$$

where λ represents the rate at which customers arrive, since the fact that $E[A_s] = \lambda s$ implies

$$\lim_{s \to \infty} \frac{A_s}{s} = \lambda$$

by the strong law of large numbers.

If $T = \{T_n;\ n = 0, 1, \ldots\}$ is the arrival time or point process on the real line so that T_n depicts the time of arrival of the nth customer, then

$$P\{T_{n+1} - T_n > t | T_0, T_1, \ldots, T_n\} = e^{-\lambda t}\ , \qquad t \geq 0.$$

Thus, $\{T_{n+1} - T_n\}$ is a sequence of independent and identically distributed random variables which have the exponential distribution with parameter λ. This relationship between the Poisson process and the exponential distribution is unique, in the sense that an arrival process is Poisson if and only if interarrival times have the exponential distribution.

2.2. Forward and Backward Recurrence Times

Another important property of the Poisson process is related with the so-called forward and backward recurrence times as depicted in Figure 1.

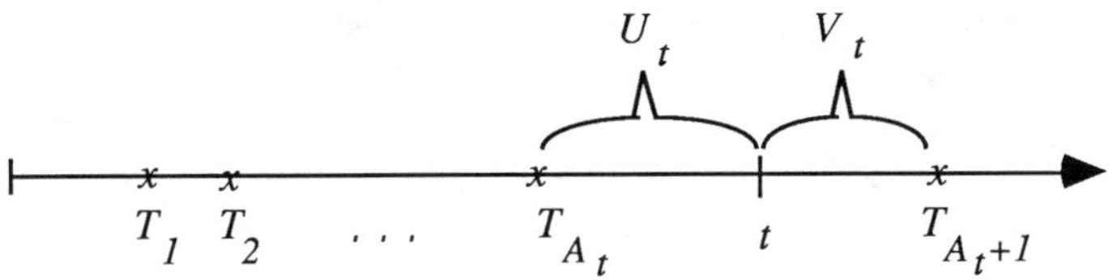

Figure 1. Forward and Backward Recurrence Times

At any time t, the forward recurrence time V_t is the time until the next arrival, i.e., $V_t = T_{A_t+1} - t$,and the backward recurrence time is the time since the last arrival, i.e., $U_t = t - T_{A_t}$. This also implies that $U_t + V_t = T_{A_t+1} - T_{A_t}$ is the length of the time interval covering the point t.

The memorylessness property of the exponential distribution implies that V_t has the exponential distribution with parameter λ, in other words

$$P\{V_t > u\} = e^{-\lambda u}, \qquad u \geq 0$$

and one can also derive the distribution

$$P\{V_t + U_t > u\} = \begin{cases} (1 + \lambda t)e^{-\lambda u}, & u \geq t \\ (1 + \lambda u)e^{-\lambda u}, & u < t \end{cases}$$

so that we arrive at the paradox

$$E[T_{A_t+1} - T_{A_t}] = E[V_t + U_t] = \frac{2 - e^{-\lambda t}}{\lambda} > \frac{1}{\lambda} = E[T_{n+1} - T_n]$$

for all $n \geq 0$, $t > 0$. As a matter of fact, $\lim_{t \to \infty} E[V_t + U_t] = \frac{2}{\lambda}$.

This result, which seems rather unintuitive at first glance, is an indication of an amazing implication of exponential interarrival times. The exponential form of the density function implies that most of the interarrival durations are rather short when compared with the mean $1/\lambda$, and those that are longer than $1/\lambda$ better be very long so that the total averages out to $1/\lambda$. Thus, the arrival times $\{T_n\}$ must form a clustured sequence of points on the real line. Therefore, any point t is more likely to fall in a longer interval, one that has expectation larger than $1/\lambda$.

2.3. Superposition and Decomposition of Poisson Processes

Another important result on Poisson processes is that the superposition and decomposition of Poisson processes yield Poisson processes. To be more precise, suppose $A^1, A^2, ..., A^n$ are independent Poisson processes representing different arrival types with respective rates $\lambda_1, \lambda_2, ..., \lambda_n$, then the process that counts all types of arrivals

$$A_t = A_t^1 + A_t^2 + \cdots + A_t^n , \qquad t \geq 0$$

is also a Poisson process with rate $\lambda = \lambda_1 + \lambda_2 + \cdots + \lambda_n$. Similarly, if A is a Poisson process with rate λ, where each arrival can be classified to be of type $1, 2, ..., n$ with respective probabilities $p_1, p_2, ..., p_n$, then the processes $A^1, A^2, ..., A^n$ that count arrivals of types $1, 2, ..., n$ respectively are all independent Poisson processes with respective rates $\lambda_1 = \lambda p_1, \lambda_2 = \lambda p_2, ..., \lambda_n = \lambda p_n$.

2.4. Compound Poisson Processes

In many applications, arrivals occur in batches, rather than singly. This violates the first sample-path property of an ordinary Poisson process. However, a small modification of property 1 as,

$1'$. $t \rightarrow A_t$ is right-continuous, and increasing by jumps only $(A_0 = 0)$,

leads to the definition of a compound Poisson process. In this case , it is quite clear that the batch arrival times $\{T_n\}$ form an ordinary Poisson process with rate λ, with the additional generalization that the consequtive batch magnitudes $\{B_n\}$ form a sequence of independent and identically distributed random variables with some common distribution F that has mean b and variance σ_b^2.

This additional feature does not really impose to many difficulties since the compound process A now has the characterization

$$A_t = \sum_{T_n \leq t} B_n .$$

Therefore, probabilistic quantities associated with A can be derived by using this characterization and conditioning on the Poisson batch arrival times $\{T_n\}$. In particular,

$$E[A_t] = \lambda b t$$
$$Var(A_t) = \lambda(\sigma_b^2 + b^2)t$$

and the distribution of A_t is

$$P\{A_t \leq x\} = \sum_{n=0}^{\infty} \frac{e^{-\lambda t}(\lambda t)^n}{n!} F^{*n}(x) , \qquad x \geq 0$$

where F^{*n} is the n-fold convolution of F, i.e., $F^{*n}(x) = P\{B_1 + B_2 + \cdots + B_n \leq x\}$.

In queueing applications, the batch sizes $\{B_n\}$ naturally take values in $\{0, 1, 2, ...\}$ with respective probabilities

$$P\{B_n = k\} = p_k , \qquad k = 0, 1, 2, ... \ .$$

Therefore, each arrival can be classified to be of type $0, 1, 2, ...$ depending on the batch size. If A_t^i represents the number of batches of size i that arrived until time t, than each A^i is an ordinary Poisson process with rate λp_i, and the compound Poisson process A is simply the superposition of the ordinary Poisson processes $\{A^i\}$. This leads to the simple characterization

$$A_t = \sum_{i=0}^{\infty} A_t^i , \qquad t \geq 0$$

which can be exploited to analyze probabilistic features of a compound Poisson process through those of ordinary Poisson processes.

3. MARKOV CHAINS

3.1. Introduction

Another well-known stochastic process that arize in the context of queueing models are Markov chains. Although the stochastic evolution of a queue is observed in continuous time, it is sometimes useful, or necessary, to observe the queue at certain discrete random points in time. This is done in order to reflect various properties of the discrete time process onto the primary continuous time process. In many cases, especially for M/G or G/M type queueing models, the observation times are chosen so that the number of customers observed to be in the system form a Markov chain. This technique, known as embedding, was introduced by Kendall (1953), and is almost invariably used in all M/G or G/M type queues. Sections 3.1 and 3.2 in Chapter 3 illustrate how this technique can be utilized.

3.2. Definition and the Markov Property

Let X_n denote the state of a system at time $n = 0, 1, ...$, where X_n generally represents the number of customers in the system at the time of the nth observation for queueing models. Assume that the process X takes values in the discrete set $E = \{0, 1, 2, ...\}$. Then, X is called a Markov chain if it satisfies the following so-called Markov property:

$$P\{X_{n+1} = j | X_0, X_1, ..., X_n\} = P\{X_{n+1} = j | X_n\}$$

for all $j \in E$, $n \geq 0$. Moreover, X is called time-homogeneous if

$$P\{X_{n+1} = j | X_n = i\} = P(i, j)$$

for all $i, j \in E$ independent of n, where P is called the transition matrix of X.

The Markov property is, in a sense, similar to the memorylessness property of the exponential distribution. If n denotes the present time, then the future of the process $\{X_{n+1}, X_{n+2}, ...\}$ is independent of its past $\{X_0, X_1, ..., X_n\}$ given the present X_n. In the context of many complex queueing models, the observation points can be deliberately chosen so that the number of customers in the system at some future time depends only on the presently observed quantity, and not on what it was in the past. In an $M/G/1$ queue, for example, the number of customers observed after service completions form a Markov chain since at the time of the observation there is no service being conducted, and predicting the time until the next service completion does not require information from the past. Similarly, the observation times that define the embedded Markov chain are taken to be customer arrival times in a $G/M/1$ queue. This technique of using embedded Markov chains enable analyst to utilize the available theory on Markov chains. The rest of this section presents a brief review of the basic results.

3.3. Transient Analysis and First Passage Times

The statistical information regarding a Markov chain X is summarized by its transition matrix P, where $P(i, j)$ is simply the probability that the chain will move from state i to state j in a single step. A simple analysis leads to the fact that n step evolutions are governed by the conditional distribution

$$P\{X_n = j | X_0 = i\} = P^n(i, j) , \qquad i, j \in E$$

where P^n is nth power of P. Thus, the n-step transition matrix P^n can be easily computed using linear algebra.

The analysis of first passage times and the total number of visits to a fixed state constitute an important part of the general theory of Markov chains. For $j \in E$, let

$$T^j = \inf\{n \geq 1 \; ; X_n = j\}$$

be the time of first visit to state j, and

$$N^j = \sum_{n=0}^{\infty} 1_j(X_n)$$

be the total number of times state j is visited. Various quantities of interest associated with these random variables are given by the matrices $F_k(i, j) = P\{T^j = k | X_0 = i\}$ for $k = 1, 2, ..., F(i, j) = P\{T^j < \infty | X_0 = i\}$, $M(i, j) = E[T^j | X_0 = i]$, and $R(i, j) =$

$E[N^j|X_0 = i]$. Using the Markov property at $n = 1$ by conditioning on X_1, one can easily verify that these matrices satisfy

$$F_k(i,j) = \begin{cases} P(i,j), & k = 1 \\ \sum_{l \neq j} P(i,l)F_{k-1}(l,j), & k \geq 2 \end{cases}$$

$$F(i,j) = P(i,j) + \sum_{l \neq j} P(i,l)F(l,j)$$

$$M(i,j) = 1 + \sum_{l \neq j} P(i,l)M(l,j)$$

$$R(i,j) = I(i,j) + \sum_{l \in E} P(i,l)R(l,j)$$

for $i, j \in E$. Thus, F_k can be computed recursively, and the other matrices satisfy systems of linear equations. The existence and uniqueness of a solution to these set of equations depend, however, on various properties of the given transition matrix P.

3.4. Classification of the States

The states of a Markov chain are classified according to how it behaves in these states. In particular , a state $i \in E$ is called:

1. Recurrent if $F(i,i) = 1$,
    1.1 Null recurrent if $F(i,i) = 1$ with $M(i,i) = \infty$,
    1.1 Non-null recurrent if $F(i,i) = 1$ with $M(i,i) < \infty$,
2. Transient if $F(i,i) < \infty$,
3. Absorbing if $P(i,i) = 1$,
4. Periodic with period δ if $\delta \geq 2$ is the largest integer for which $P\{T^i = n\delta$ for some $n \geq 1|X_0 = i\} = 1$,
5. Aperiodic if the largest integer δ for which

$$P\{T^i = n\delta \text{ for some } n \geq 1|X_0 = i\} = 1$$

   is 1,
6. Ergodic if it is non-null recurrent, aperiodic, and it can be reached from all of the other states, i.e., $F(j,i) > 0$ for all $j \in E$.

The description of these classes is provided by their names. Recurrent, for example, means that the state recurs itself so that the probability of ever reaching a recurrent state from itself is 1, i.e., $F(i,i) = 1$. For a transient state i, on the other hand, there is always a positive immediate probability $1 - F(i,i) > 0$ that the process escapes, never to return back to i.

3.5. Ergodic and Potential Theory

A Markov chain is called ergodic if all states are ergodic. Ergodic Markov chains play an important role in queueing theory. This is due to the fact that ergodicity implies the existence of the limiting or steady-state distribution of the Markov chain. Moreover, time-averages associated with the chain equal to the corresponding state-averages which can be computed once the steady-state distribution is determined.

Ergodic theory on Markov chain state that the steady-state distribution

$$\pi(j) = \lim_{n\to\infty} P^n(i,j) = \lim_{n\to\infty} P\{X_n = j | X_0 = i\}, \qquad i, j \in E$$

of an ergodic Markov chain exists, and is the unique solution of the system of linear equations

$$\pi(j) = \sum_{i\in E} \pi(i)P(i,j), \qquad j \in E.$$

with $\sum_i \pi(i) = 1$. Moreover,

$$\lim_{n\to\infty} \frac{1}{n+1} \sum_{k=0}^{n} g(X_k) = \sum_{i\in E} \pi(i)g(i)$$

so that time-averages can be easily computed using π. Once can think of g as a reward vector so that $\sum_i \pi(i)g(i)$ is the average reward earned by the process X in one period in the long run. In the context of queueing theory, where $\{X_n\}$ generally represents the number of customers in the system at certain points in time, ergodicity implies stability for the queueing system. Thus, stability conditions can be obtained through the system of equations, $\pi = \pi P$ with $\pi 1 = 1$.

Similarly, linear algebra plays the key computational role in potential theory. The main result in this area states that

$$f(i) = E[\sum_{n=0}^{\infty} \alpha^n g(X_n) | X_0 = i], \qquad i \in E$$

is the unique solution of the system of linear equations

$$f(i) = g(i) + \alpha P f(i), \qquad i \in E$$

for any $\alpha \in [0,1)$. Here, α represents a periodic discount factor, and once again, g can be viewed as a reward vector, so that $f(i)$ is the expected total discounted reward earned by the process X in its lifetime given that its initial state is i. An analysis on $f = g + \alpha P f$ yields the characterization that $f = R^\alpha g$ where $R^\alpha = (I - \alpha P)^{-1}$ is the α-potential matrix of X defined by $R^\alpha = \sum_n \alpha^n P^n$.

4. MARKOV PROCESSES

4.1. Introduction

The fundamental stochastic process associated with any queueing system is one which describes the stochastic evolution of the number of customers in the system. In the case of a single queue, such a process will take values in the discrete set $E = \{0, 1, 2, ...\}$, and it will either increase or decrease by jumps due to customer arrivals or departures after service completions respectively. In the so-called M/M type queueing models, or Markov models, this process constitutes a Markov process with a discrete state space. These models cover a substantial part of the existing literature due mainly to their computational tractability, as well as applicability to real-life problems.

4.2. Definition and the Markov Property

Let Y_t denote the state of a system at time $t \geq 0$, where Y_t generally represents the number of customers present at time $t \geq 0$ for queueing models. Then, Y is called a Markov process if it satisfies the following so-called Markov property

$$P\{Y_{t+s} = j | Y_u;\ u \leq t\} = P\{Y_{t+s} = j | Y_t\}$$

for all $j \in E$, $t, s \geq 0$. Moreover, Y is called time-homogeneous if

$$P\{Y_{t+s} = j | Y_t = i\} = P_s(i, j)$$

for all $i, j \in E$, $s \geq 0$ independent of t, where (P_s) is called the transition function of Y.

Note that this Markov process Y is simply the continuous-time version of the Markov chain X of the previous section. This is why they are sometimes referred to as continuous-time Markov chains. Once again, the Markov property simply states that at any present time t, the future of the process $\{Y_{t+s};\ s \geq t\}$ is independent of its past $\{Y_u;\ u \leq t\}$ given the present state Y_t. In the context of queueing models, the Markov property implies that, at any time, the amount of time that must be waited until the next customer arrival or service completion must both have the exponential distribution. This is due to the memorylessness property of the exponential distribution, and no information from the past is required to predict the times of future events regarding arrivals or departures.

4.3. Stochastic Structure of a Markov Process

The stochastic structure of Y can be characterized by a Markov chain, and exponentially distributed sojourns, as depicted in Figure 2 where $\{X_n\}$ represents the consecutive states visited by Y, and $\{T_n\}$ denotes the jump times of Y. In other words,

$$T_0 = 0 , \qquad T_{n+1} = \inf\{t > T_n :\ Y_t \neq Y_{T_n}\} , \qquad n \geq 1$$

and $X_n = Y_{T_n}$. For technical reasons, we assume that Y is regular, i.e., $t \to Y_t$ is right-continuous and $\sup_n T_n = \infty$.

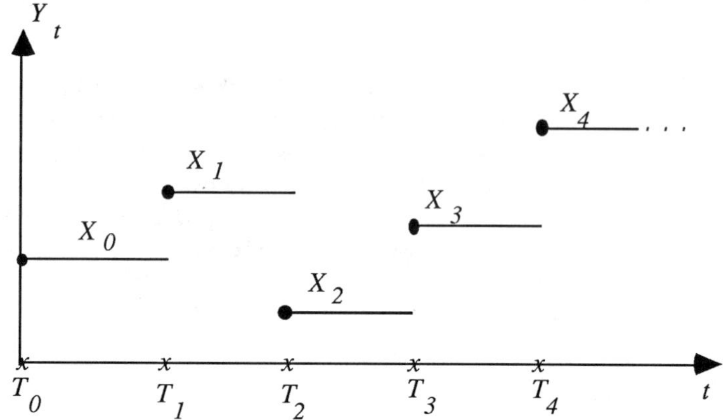

Figure 2. A Typical Sample-Path of the Markov Process Y

The stochastic characterization of Y can be made by the following two results:

1. X is a Markov chain with some transition matrix P,
2. Given $\{X_n\}$, the random variables $\{T_{n+1} - T_n\}$ are independent and exponential distributions with parameters $\{\lambda(X_n)\}$ for some jump rate vector λ.

Therefore, the sequence of states visited by Y form a Markov chain, and the amount of time spent in each state has the exponential distribution. If the process is in state i, than the next state visited will be j with probability $P(i,j)$ after a random duration which has the exponential distribution with parameter $\lambda(i)$. The statistical properties of Y are summarized in following so-called generator matrix

$$A(i,j) = \begin{cases} -\lambda(i), & j = i \\ \lambda(i)P(i,j), & j \neq i \end{cases}$$

for $i,j \in E$ where $A(i,j)$ is the rate at which the process jumps from state i to state j for $i \neq j$. We suppose without loss of generality that $P(i,i) = 1$ whenever $\lambda(i) = 0$ for any $i \in E$, so that state i is absorbing for X. Using the Markov property, one can easily obtain the Chapman-Kolmogrov forward and backward equation

$$\frac{d}{dt}P_t = AP_t = P_t A , \qquad t \geq 0$$

so that $A = (dP_t/dt)\big|_{t=0}$ is the derivative of the transition function P_t at $t = 0$ since $P_0 = I$. Moreover, the solution is

$$P_t(i,j) = \sum_{n=0}^{\infty} \frac{t^n}{n!} A^n(i,j) \qquad i,j \in E, t \geq 0$$

and this justifies why A is called the generator of Y. The infinitesimal behaviour of P_t in the vicinity of $t = 0$ generates P_t for all $t \geq 0$. This is the main result concerning the transient behaviour of Y.

The classification of the states of Y is made with respect to the classification of the states of the embedded chain X, as described in the previous section. Note that the number of times a fixed state j is entered by Y is equal to the number of visits to j by X. Therefore, the probability of ever visiting j from i, and the expected number of entrances to j starting at i are still given by $F(i,j)$ and $R(i,j)$ respectively. Similarly, various probabilistic quantities of interest regarding Y can be computed through those of X by exploiting the stochastic characterization described in Figure 2.

4.4. Ergodic and Potential Theory

The Markov process is called ergodic if the embedded chain X is ergodic. In this case, the steady-state distribution

$$\mu(j) = \lim_{t \to \infty} P_t(i,j) = \lim_{t \to \infty} P\{Y_t = j | Y_0 = i\} , \qquad i,j \in E$$

exists independent of i, and is the unique solution of the system of linear equations

$$\sum_{i \in E} \mu(i) A(i,j) = 0 , \qquad j \in E$$

with $\sum_i \mu(i) = 1$. If the process Y earns reward at rate $g(Y_s)$ at time s for some reward vector g, then the average reward can be computed as

$$\lim_{t \to \infty} \frac{1}{t} \int_0^t g(Y_s) ds = \sum_{i \in E} \mu(i) g(i) .$$

In queueing models where Y represents the number of customers in the system, the ergodicity condition, $\mu A = 0$ with $\mu 1 = 1$, yields the stability condition for the queue. Another well-known result in ergodic theory is that

$$\mu(j) = \frac{\pi(j)/\lambda(j)}{\sum_{i \in E} \pi(i)/\lambda(i)} , \qquad j \in E$$

where π is the steady-state distribution of X; i.e., π is the unique solution of $\pi = \pi P$ with $\pi 1 = 1$.

For any reward rate vector g, the expected total discounted earning given that the initial state is i, i.e.,

$$f(i) = E\left[\int_0^\infty e^{-\alpha t} g(Y_t) dt \Big| Y_0 = i \right] , \qquad i \in E$$

is the unique solution of the system of linear equations

$$\alpha f(i) = g(i) + A f(i) , \qquad i \in E$$

for any continuous discount factor $\alpha > 0$. A further analysis yields that f has the characterization $f = U^\alpha g$ where $U^\alpha = (\alpha I - A)^{-1}$ is the α-potential matrix of Y defined by $U^\alpha = \int_0^\infty e^{-\alpha t} P_t dt$.

4.5. Markov Chains Subordinated to Poisson Processes

A simpler stochastic characterization of Y can be obtained in the case $\sup_i \lambda(i) = c < \infty$. The transient behaviour of Y is now described through

$$P_t(i,j) = \sum_{n=0}^{\infty} \frac{e^{-ct}(ct)^n}{n!} K^n(i,j) , \qquad i,j \in E, \ t \geq 0$$

where $K = I + \frac{1}{c} A$ is a Markov transition matrix. This implies that Y behaves like a Markov chain Z with transition matrix K subordinated to an ordinary Poisson process N with rate c. In other words,

$$P\{Y_t = j | Y_0 = i\} = P\{Z_{N_t} = j | Z_0 = i\} , \qquad i,j \in E, \ t \geq 0 .$$

It is clear that $\{Z_{N_t}\}$ is a stochastic process that changes state only at the arrival times of the Poisson process N, and when it does, it does so in accordance with the Markov chain Z. Thus, Y can be analyzed using this characterization through N and Z. In particular, it follows that the steady-state distribution μ of Y is also the steady-state distribution of Z, and μ is the unique solution of $\mu = \mu K$ with $\mu 1 = 1$. Moreover, $(\alpha + c)U^\alpha = \sum_n (c/(\alpha+c))^n K^n$ so that the α-potential matrix of Y is simply a multiple of the $[c/(\alpha + c)]$-potential matrix of Z.

As a special case when $\alpha = 0$, the expected amount of time spent in state j by Y starting at the initial state i is $U(i,j) = \frac{1}{c} \sum_n K^n(i,j)$, which is simply the expected number of times Z visits j from i, $\sum_n K^n(i,j)$, times the expected duration of each visit, $1/c$, since sojourns are now exponential with parameter c. By similar reasoning, it follows that $U(i,j) = R(i,j)/\lambda(j)$ where R is the potential matrix of the embedded Markov chain X.

4.6. Birth and Death Processes

Markov processes which either increase or decrease by one unit whenever there is a state change are of utmost importance in queueing theory. These are referred to as birth and death processes where birth implies a customer arrival, whereas death implies a customer departure. Given that there are i customers in the system, if the time until the next arrival is exponentially distributed with rate a_i, and the time until the next departure is exponentially distributed with rate b_i, then Y is a Markov process with generator

$$
A = \begin{array}{c} \\ 0 \\ 1 \\ 2 \\ \cdot \\ \cdot \\ \cdot \end{array}
\begin{array}{cccc}
0 \quad\quad & 1 \quad\quad & 2 \quad\quad & \cdot \;\; \cdot \;\; \cdot \\
\left(\begin{array}{cccc}
-a_0 & a_0 & & \\
b_1 & -(a_1 + b_1) & a_1 & \\
0 & b_2 & -(a_2 + b_2) & a_2 \\
\cdot & \cdot & \cdot & \cdot \\
\cdot & \cdot & \cdot & \cdot \\
\cdot & \cdot & \cdot & \cdot
\end{array}\right)
\end{array}
$$

In our characterization, the transition matrix of the embedded Markov chain is

$$
P(i,j) = \begin{cases} a_i/(a_i + b_i), & j = i + 1 \\ b_i/(a_i + b_i), & j = i - 1 \end{cases}
$$

and $\lambda(i) = a_i + b_i$ for any $i \in E$.

The steady-state distributions exists if $\mu A = 0$ with $\mu 1 = 1$ has a unique solution. The stability condition for the queue is thus found to be

$$
d = 1 + \sum_{i=1}^{\infty} \frac{a_0 a_1 \cdots a_{i-1}}{b_1 b_2 \cdots b_i} < \infty \,,
$$

in which case the steady-state distribution is

$$
\mu(i) = \begin{cases} 1/d, & j = 0 \\ (1/d)\frac{a_0 a_1 \cdots a_{j-1}}{b_1 b_2 \cdots b_j}, & j \geq 1. \end{cases}
$$

As we shall see in more detail in Chapter 2, various special cases can be analyzed. In an $M/M/1$ queue, for example, it suffices to take $a_0 = a_1 = \cdots = a$ and $b_1 = b_2 = \cdots = b$ to find the stability condition $\rho = a/b < 1$, and steady-state distribution $\mu(j) = (1 - \rho)\rho^j$, $j = 0, 1, \ldots$. For an $M/M/\infty$ queue, on the other hand, $a_0 = a_1 = \cdots = a$ and $b_i = ib$ so that the stability condition, $d = e^\rho < \infty$, is always satisfied to arrive at the steady-state distribution $\mu(j) = \rho^j e^{-\rho}/j!$, $j = 0, 1, \ldots$. Other queueing models which can be analyzed using birth and death processes are $M/M/s$ queues with identical and non-identical servers, $M/M/s$ queues with finite system capacity, and $M/M/s$ queues with finite population size. These examples, as well as the results stated in this subsection demonstrate that M/M type queueing systems can be modeled by Markov processes which possess nice computationally tractable procedures regarding ergodic and potential analysis.

5. RENEWAL PROCESSES

5.1. Introduction

One of the very powerful tools in ergodic analysis is provided by renewal theory. This necessitates the use of renewal processes which typically represent an increasing sequence of random points in time, when certain events that interest us occur. This could be, for example, customer arrival times where interarrival times are no longer restricted to be exponentially distributed as in Poisson processes. In queueing systems, the consecutive times at which busy cycles start generally constitute a renewal process. The importance of renewal processes is not due to their rather simple stochastic structure, but more due to their association with renewal theory and regenerative processes, as we shall demonstrate in this section.

5.2. Definition and Basic Results

The stochastic process $S = \{S_n;\ n = 0, 1, ...\}$ is called a renewal process provided that $S_0 = 0$ and $S_{n+1} = S_n + U_{n+1}$, $n > 0$, where $\{U_n\}$ is a sequence of independent and identically distributed random variables with common distribution F. We call S_n the time of the nth renewal, and define $N_t = \sum_n 1_{\{S_n \leq t\}}$ to be the total number of renewals, including that at time zero, until time t. It should be noted that N is a generilization of Poisson processes where interarrival times, or interrenewal times $\{U_n\}$, are now independent but arbitrarily distributed. Obviously, if F is the exponential distribution, then N is a Poisson process.

It is clear that $S_n = U_1 + U_2 + \cdots + U_n$ for all $n \geq 0$ so that

$$P\{S_n \leq t\} = F^n(t) , \qquad t \geq 0$$

where F^n is the n-fold convolution of F with itself. Moreover, the distribution of N_t is

$$P\{N_t = k\} = F^{k-1}(t) - F^k(t) , \qquad k \geq 1,\ t \geq 0$$

since $\{N_t = k\} = \{S_{k-1} \leq t < S_k\}$. The expectation

$$R(t) = E[N_t] = \sum_{n=0}^{\infty} P\{S_n \leq t\} = \sum_{n=0}^{\infty} F^n(t) , \qquad t \geq 0$$

is called the renewal function corresponding to F. It is clear that R is right-continuous, increasing, and finite with the assumption that $F(0) < 1$ so that trivialities are excluded.

The renewal function plays a similar role as the potential matrix in Markov chains and processes. If the nth renewal at time S_n earns reward $g(S_n)$ where g is a bounded reward function that vanishes outside a finite interval, then

$$E[\sum_{n=0}^{\infty} f(S_n)] = \int_{[0,\infty)} R(du)f(u) .$$

The renewal process S is called recurrent if $F(\infty) = P\{U_n < \infty\} = 1$, otherwise it is called transient. The implications of these definitions are self-explanatory. In particular, transience implies that at any renewal time there is always a positive probability, $1 - F(\infty)$, that no more renewals will be observed. In this case, it follows that

$$P\{\lim_{t \to \infty} N_t = k\} = (1 - F(\infty))F(\infty)^{k-1} , \qquad k = 1, 2, \ldots$$

and $R(\infty) = \lim_{t \to \infty} R(t) = E[\lim_{t \to \infty} N_t] = 1/(1 - F(\infty)) < \infty$. Moreover, the time of the last renewal $L = \sup_n \{S_n : \; S_n < \infty\}$, or the so-called lifetime of the process, is almost surely finite with distribution

$$P\{L \le t\} = (1 - F(\infty))R(t) , \qquad t \ge 0 .$$

and expectation

$$E[L] = \int_0^n (F(\infty) - F(t))dt/(1 - F(\infty)) .$$

For a recurrent renewal process, on the other hand, it is clear that $\lim_t N_t = \lim_t R(t) = L = \infty$. However, the time-averages

$$\lim_{t \to \infty} \frac{N_t}{t} = \lim_{t \to \infty} \frac{R(t)}{t} = \frac{1}{m}$$

exist where $m = E[U_n]$ is the mean interrenewal duration so that $1/m$ is the rate at which renewals occur. Note that if N is a Poisson process with rate λ, then $1/m = \lambda$ and $R(t) = 1 + \lambda t$.

5.3. Renewal Theory

In dealing with renewal processes, one often obtains a functional equation of the form

$$f(t) = g(t) + \int_{[0,t)} F(du)f(t - u) , \qquad t \ge 0$$

called the renewal equation. This equation is often abbreviated as $f = g + F * f$ in compact form where $*$ represents the convolution operation defined by the second term on the right-hand side. Renewal theory, which is the study of the renewal equation, states that the renewal equation has a unique solution given by

$$f(t) = R * g(t) = \int_{[0,t)} R(ds)g(t - s) , \qquad t \ge 0$$

where $R = \sum_n F^n$ is the renewal function corresponding to F.

Although the solution $R*g$ exists in theory, it is very difficult to compute in practice due to the difficulty in determining the renewal function R. The limiting behaviour of the solution, however, can be easily determined by the fact that

$$\lim_{t \to \infty} R * g(t) = R(\infty)g(\infty) = \frac{g(\infty)}{1 - F(\infty)}$$

if $F(\infty) < 1$ (i.e., the transient case) provided that $g(\infty) = \lim_{t \to \infty} g(t)$ exists, and

$$\lim_{t \to \infty} R * g(t) = \frac{1}{m} \int_0^\infty g(y)dy$$

if $F(\infty) = 1$ (i.e., the recurrent case) provided that g is directly Riemann integrable. The reader is referred to Çınlar (1975, p.295) for the definition of directly Riemann integrablity and related results. Nice behaving functions, like positive and continuous function that vanish outside finite intervals, are directly Riemann integrable; so are ordinary Riemann integrable functions which are positive and decreasing.

5.4. Regenerative Processes

The primary application of renewal theory is in the analysis of regenerative processes. A process $Z = \{Z_t; \ t \geq 0\}$ is said to be regenerative provided that there exists a sequence $\{S_n\}$ of stopping times of Z such that

1. S is a renewal process,
2. $P\{Z_{S_n+t} \in B | Z_u; \ u \leq S_n\} = P\{Z_t \in B\} , \qquad t \geq 0,$
   for any $n \geq 0$ and B in the state space E of Z.

This definition of a regenerative process implies that the process regenerates itself at the renewal times so that the future of the process is a probabilistic replica at each renewal time. This is why $\{S_n\}$ are also referred to as regeneration times.

Defining $f(t) = P\{Z_t \in B\}$, the renewal theoretic argument made by conditioning on the first renewal time S_1

$$f(t) = P\{Z_t \in B, \ S_1 > t\} + P\{Z_t \in B, \ S_1 \leq t\} , \qquad t \geq 0$$

leads to the renewal equation

$$f(t) = g(t, B) + \int_{[0,t)} F(ds)f(t - s) , \qquad t \geq 0$$

where $g(t, B) = P\{Z_t \in B, \ S_1 > t\}$ and F is the interrenewal distribution of s S. Therefore, the transient behaviour of Z is characterized by

$$P\{Z_t \in B\} = \int_{[0,t)} R(ds)g(t - s, B) , \qquad t \geq 0$$

through renewal theory. Moreover, the limiting of Z is

$$\lim_{t \to \infty} P\{Z_t \in A\} = \frac{1}{m} \int_0^\infty g(u, B) du$$

if S is recurrent with $m < \infty$, provided that $t \to g(t, B)$ is Riemann integrable. If S' is transient, or recurrent with $m = \infty$, then $\lim_{t \to \infty} P\{Z_t \in A\} = 0$ trivially.

As an illustration, consider the forward recurrence time process V of a renewal process S defined by

$$V_t = S_{N_t} - t , \qquad t \geq 0$$

so that V_t is the amount of time that must be waited after t until the next renewal. The sample-path structure of V is pictured in Figure 3.

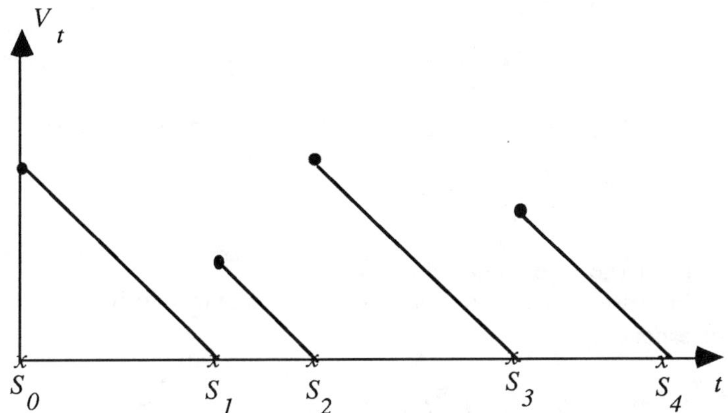

Figure 3. A Typical Sample-Path of the Forward Recurrence Time Process

It is clear that V is a regenerative process, and taking $B = [y, \infty)$, one obtains

$$g(t, B) = P\{V_t > y, \ S_1 > t\} = P\{S_1 > t + y\} = 1 - F(t + y) .$$

Thus, the transient distribution is

$$P\{V_t > y\} = \int_{[0,t)} R(du)(1 - F(t + y - u)) , \qquad y \geq 0, \ t \geq 0$$

and the limiting distribution is

$$\lim_{t \to \infty} P\{V_t > y\} = \frac{1}{m} \int_0^\infty (1 - F(t + y)) dt ,$$

$$= \frac{1}{m} \int_y^\infty (1 - F(x)) dx , \qquad y \geq 0$$

which has mean $\lim_{t\to\infty} E[V_t] = (m^2 + v^2)/2m$ where v^2 is the variance of F. The limiting distribution of V is also referred to as the stationary forward recurrence time distribution of F. Recall from Section 2 that for a Poisson process, where the interrenewal distribution F is exponential, the stationary forward recurrence time distribution is still exponential with the same parameter. As a matter of fact, the exponential distribution is the only distribution for which this identicality is true.

5.5. Delayed and Stationary Renewal Processes

The process $\hat{S} = \{\hat{S}_n;\ n \geq 0\}$ is called a delayed renewal process if $\hat{S}_0 \geq 0$ and $S = \{\hat{S}_n - \hat{S}_0;\ n \geq 0\}$ a renewal process independent of $\hat{S}_0$. Here $\hat{S}_0$ is called the delay with some arbitrary distribution G, and the renewal process S starts after this delay. If F is the interrenewal distribution of S with renewal function R, then the renewal function $\hat{R}$ corresponding to $\hat{S}$ satisfies

$$\hat{R}(t) = G * R(t) , \qquad t \geq 0$$

so that $\hat{R}(\infty)/(1 - F(\infty))$ whenever $F(\infty) < 1$ (i.e., the transient case), and

$$\lim_{t\to\infty} \hat{R}(t)/t = G(\infty)/m$$

whenever $F(\infty) = 1$ (i.e., the recurrent case).

An interesting case is obtained when G is the stationary forward recurrence time distribution corresponding to F, i.e.,

$$G(y) = 1 - \frac{1}{m} \int_y^\infty (1 - F(x))dx , \qquad y \geq 0 .$$

It is now possible to show that $\hat{R}(t) = (1/m)t,\ t \geq 0$, which implies that the renewal counting process of $\hat{S}$ behaves similar to an ordinary Poisson process with rate $1/m$ in expectation. A delayed renewal process satisfying this property is called a stationary renewal process. This is actually the process observed by an individual who arrives at time infinity, to start observing an already ongoing ordinary renewal process.

6. MARKOV RENEWAL PROCESSES

6.1. Introduction

Markov renewal processes are, in a sense, combinations of Markov chains and renewal processes which behave like a Markov chain when state changes occurs, and like a renewal process with respect to the timing of these state changes. In this regard, they do not only generalize all of the processes studied in this Chapter, but provide more elegant stochastic modeling tools and analytical techniques as well. They can be used in many fields of science and engineering including, of course, queueing theory. Their application in queueing theory generally arises in the context of non M/M, M/G or G/M type queueing models. Sections 3.2 and 3.3 illustrate their applicability in more detail. As in ordinary renewal processes, the association of Markov renewal processes with Markov renewal theory and semi-regenerative processes make them more appealing to theoreticians as well as practitioners.

6.2. Definition and Basic Results

The stochastic process $(X, T) = \{(X_n, T_n); \ n \geq 0\}$, where X takes values in a countable set E and $0 \leq T_0 \leq T_1 \leq \cdots$, is called a Markov renewal process if

$$P\{X_{n+1} = j, \ T_{n+1} - T_n \leq t | (X_k, T_k); \ k \leq n\} = P\{X_{n+1} = j, \ T_{n+1} - T_n \leq t | X_n\}$$

for all $j \in E$, $n \geq 0$, $t \geq 0$. Moreover, (X, T) is called time-homogeneous if

$$P\{X_{n+1} = j, \ T_{n+1} - T_n \leq t | X_n = i\} = Q(i, j, t)$$

for all $i, j \in E$, $t \geq 0$, independent of n, where Q is called the semi-Markov kernel of (X, T) over E.

The stochastic structure of (X, T) can be classified by the following two results:

1. X is a Markov chain with transition matrix

$$P(i, j) = \lim_{t \to \infty} Q(i, j, t) \,, \qquad i, j \in E \,,$$

2. Given $\{X_n\}$, the random variables $\{T_{n+1} - T_n\}$ are independent and have distributions $\{G(X_n, X_{n+1}, \cdot)\}$ where the distributions

$$G(i, j, t) = P\{T_{n+1} - T_n \leq t | X_n = i, \ X_{n+1} = j\} \,, \qquad t \geq 0, \ i, j \in E$$

satisfy

$$G(i, j, t) = Q(i, j, t) / P(i, j) \,, \qquad t \geq 0, \ i, j \in E$$

whenever $P(i, j) > 0$.

It is clear that if E consists of a single point, then T is a renewal process with interrenewal distribution G. Similarly, for any fixed state $j \in E$, the successive values of T_n for which $X_n = j$ form a possibly delayed renewal process. These observations and the fact that X is a Markov chain, somewhat, justifies why (X, T) is called a Markov renewal process. For technical reasons, we assume throughout this section that $\sup_n T_n = \infty$.

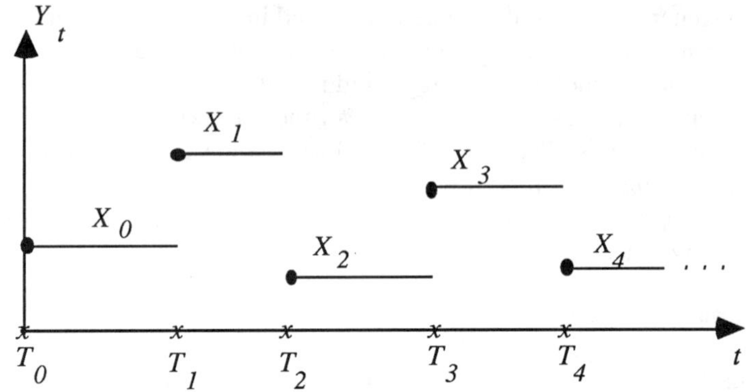

Figure 4. The Relationship Between (X, T) and Y

In visualizing a Markov renewal process, it is useful to think of the process Y defined by

$$Y_t = X_n , \quad \text{if } T_n \leq t < T_{n+1} , \qquad t \geq 0$$

so that Y_t denotes the state of the system described by X whenever $T_n \leq t < T_{n+1}$. Figure 4 describes the relationship between (X, T) and Y. Note that the two processes define one another uniquely, and Y is called the minimal semi-Markov process associated with (X, T).

The stochastic process (X, T), or Y, depicts the stochastic evolution of a system which changes state according to a Markov chain. However, the amount of time spent in these states are arbitrarily distributed depending on the state currently visited and the next state that the process will visit. Various special cases can be obtained by supposing different forms for the sojourn distributions. If, for example, $G(i, j, t) = 1 - e^{-\lambda t}$ independent of $i, j \in E$, then T forms a Poisson process with rate λ on the real line and, thus, Y is a Markov chain subordinated to this Poisson process. The Markov process of Section 4 is obtained by taking $G(i, j, t) = 1 - e^{-\lambda(i)t}$ independent of $j \in E$. It is therefore clear that the generalization provided by the Markov renewal process (X, T) concerns the sojourns in consecutive states visited; they now depend on both the present and next state with an arbitrary, not necessarily exponential, distribution. Note also that Y satisfies the Markov property at each T_n, thus it is a semi-Markov process.

The transient behaviour of (X, T) is summarized in the representation

$$Q^n(i, j, t) = P\{X_n = j, \ T_n \leq t | X_0 = i\} , \qquad i, j \in E, \ t \geq 0$$

for all $n \geq 0$, where Q^n is the n-fold convolution of Q by itself defined through

$$Q^{n+1}(i, k, t) = \sum_{j \in E} \int_{[0,t)} Q(i, j, ds) Q^n(j, k, t - s) , \qquad i, j \in E, \ t \geq 0$$

with $Q^0(i, j, t) = I(i, j)$.

6.3. Markov Renewal Function

For any fixed state $j \in E$, the consecutive points in time at which state j is entered, i.e.,

$$\hat{S}_0^j = \inf_m \{T_m; \ X_m = j\}$$

$$\hat{S}_n^j = \inf_m \{T_m > \hat{S}_{n-1}^j; \ X_m = j\} , \qquad n \geq 1$$

form a possibly delayed renewal process $\hat{S}^j$. The delaying occurs on $\{X_0 = i\}$ for some $i \neq j$. The renewal function corresponding to $\hat{S}^j$ on $\{X_0 = i\}$ is

$$R(i,j,t) = E[\sum_{n=0}^{\infty} I_{\{\hat{S}_n^j \leq t\}} | X_0 = i]$$

$$= E[\sum_{n=0}^{\infty} I_{\{T_n \leq t, \ X_n = j\}} | X_0 = i]$$

$$= \sum_{n=0}^{\infty} P\{X_n = j, \ T_n \leq t | X_0 = i\}$$

$$= \sum_{n=0}^{\infty} Q^n(i,j,t)$$

for any $i, j \in E$, $t \geq 0$. Defining

$$F(i,j,t) = P\{\hat{S}_0^j \leq t | X_0 = i\} , \qquad i, j \in E, \ t \geq 0 ,$$

it follows that $t \to R(i,j,t)$ is a renewal function with delay distribution $F(i,j,\cdot)$ and interrenewal distribution $F(j,j,\cdot)$. Note that $F(i,j,\infty) = F(i,j)$ is the probability of ever reaching state j from state i, given by the F matrix of the Markov chain X which was characterized by a system of linear equation in Section 3.

If state j is transient for the Markov chain X, then $\hat{S}^j$ is a transient Markov renewal process since $F(j,j,\infty) = F(j,j) < 1$. The results presented in the previous section on renewal processes state that $R(i,j,\infty) = F(i,j,\infty)[1/(1 - F(j,j,\infty))] = F(i,j)/(1 - F(j,j))$. On the other hand, if state j is recurrent, i.e., $F(j,j) = 1$, then $\lim_{t \to \infty} R(i,j,t) = \infty$ if $F(i,j) > 0$, but

$$\lim_{t \to \infty} \frac{R(i,j,t)}{t} = \frac{F(i,j)}{\eta(j)}$$

where $\eta(j) = E[S_1^j | X_0 = j]$ is the mean interrenewal duration.

6.4. Markov Renewal Theory

In almost any stochastic model where a Markov renewal process (X, T) is used, one often encounters the functional equation

$$f(i,t) = g(i,t) + \sum_{j \in E} \int_{[0,t)} Q(i,j,du) f(j, t-u) , \qquad i \in E, \ t \geq 0$$

called the Markov renewal equation. This is abbreviated as $f = g + Q * f$ where $*$ stands for the convolution operation defined by the second term on the right-hand side. The Markov renewal equation has a unique solution given by

$$f(i,t) = R * g(i,t) = \sum_{j \in E} \int_{[0,t)} R(i,j,du) g(j, t-u) , \qquad j \in E, \ t \geq 0$$

where $R = \sum_n Q^n$ is the Markov renewal kernel corresponding to Q. The existence of a unique solution is due to our assumption that $\sup_n T_n = \infty$. Otherwise, Markov renewal theory states that the solution is of the form $f = R * g + h$ where h satisfies $h = Q * h$. If $\sup_n T_n = \infty$, or E is finite, or all states of X are recurrent, then the only solution of $h = Q * h$ is $h = 0$.

The limiting behaviour of the solution $R * g$ is given by

$$\lim_{t \to \infty} R * g(i,t) = \frac{1}{\sum_{j \in E} \pi(j) m(j)} \sum_{j \in E} \pi(j) \int_0^\infty g(j, y) dy$$

where $m(j) = E[T_1 | X_0 = j]$ is the mean sojourn time in state j, provided that X is an ergodic Markov chain with steady-state distribution π and g is directly integrable with respect to π. The reader is referred to Çınlar (1975, p.332) for the definition of direct integrability with respect to a distribution, and related results.

6.5. Semi-Regenerative Processes

The characterizations on the solution of the Markov renewal equation $f = g + Q * f$ can be used to analyze semi-regenerative processes. A process $Z = \{Z_t; \ t \geq 0\}$ is said to be semi-regenerative provided that there exists a Markov renewal process (X, T) with infinite lifetime satisfying the following for all $n \geq 0$:

1. T_n is a stopping time for Z,
2. X_n is determined by $\{Z_u; \ u \leq T_n\}$,
3. If $X_n = j$, then

$$P\{Z_{T_n + t} \in B | (Z_u; \ u \leq T_n), \ X_0 = i\} = P\{Z_t \in B | X_0 = j\}$$

for any set B in the state space of Z.

Note that if the state space E of X consists of a single point, then T is a renewal process and Z is regenerative. This definition also implies that the semi-regenerative process Z is regenerative where the regeneration points are given by the renewal process $\hat{S}^j$ on $\{X_0 = j\}$ for any $j \in E$. Thus, at each renewal time $\hat{S}_n^j$, the future of the process is a probabilistic replica. However, this is not true at each Markov renewal time T_n. It is semi-regenerative in the sense that the regeneration depends on the state X_n at T_n as well.

Defining $f(i,t) = P\{Z_t \in B | X_0 = i\}$, the renewal theoretic argument made by conditioning on T_1

$$f(i,t) = P\{Z_t \in B,\ T_1 > t | X_0 = i\} + P\{Z_t \in B,\ T_1 \leq t | X_0 = i\}\ , \qquad i \in E,\ t \geq 0$$

leads to the Markov renewal equation

$$f(i,t) = g(i,t,B) + \sum_{j \in E} \int_{[0,t)} Q(i,j,ds) f(j, t-s)\ , \qquad i \in E,\ t \geq 0$$

where $g(i,t,B) = P\{Z_t \in B,\ T_1 > t | X_0 = i\}$ and Q is the semi-Markov kernel of (X,T). The transient distribution of Z is therefore given by

$$P\{Z_t \in B | X_0 = i\} = \sum_{j \in E} \int_{[0,t)} R(i,j,ds) g(j, t-s, B)\ , \qquad i \in E,\ t \geq 0\ .$$

Moreover, the limiting distribution of Z is

$$\lim_{t \to \infty} P\{Z_t \in B | X_0 = i\} = \frac{1}{\sum_{j \in E} \pi(j) m(j)} \sum_{j \in E} \pi(j) \int_0^\infty g(j,y,B)\,dy$$

indepent of $i \in E$, provided that X is ergodic with steady-state distribution π, and $t \to g(j,t,B)$ is Riemann integrable for any $j \in E$.

Note from Figure 4 that the minimal semi-Markov process Y of a Markov renewal process (X,T) is trivially semi-regenerative. For $B = \{k\}$, we obtain

$$\begin{aligned}
g(i,t,\{k\}) &= P\{Y_t = k,\ T_1 > t | X_0 = i\} \\
&= I(i,k) P\{T_1 > t | X_0 = i\} \\
&= I(i,k)\Big(1 - \sum_{l \in E} Q(i,l,t)\Big)
\end{aligned}$$

and the transient distribution is

$$\begin{aligned}
P\{Y_t = k | X_0 = i\} &= \sum_{j \in E} \int_{[0,t)} R(i,j,ds) I(j,k)\Big(1 - \sum_{l \in E} Q(j,l,t-s)\Big) \\
&= \int_{[0,t)} R(i,k,ds)\Big(1 - \sum_{l \in E} Q(k,l,t-s)\Big)\ , \qquad i,k \in E,\ t \geq 0.
\end{aligned}$$

The limiting distribution is

$$\lim_{t\to\infty} P\{Y_t = k | X_0 = i\} = \frac{\sum_{j\in E} \pi(j) \int_0^\infty I(j,k)(1 - \sum_{l\in E} Q(j,l,y))dy}{\sum_{j\in E} \pi(j)m(j)}$$

$$= \frac{\pi(k)m(k)}{\sum_{j\in E} \pi(j)m(j)} , \qquad k \in E$$

since $m(k) = E[T_1 | X_0 = k] = \int_0^\infty P\{T_1 > t | X_0 = k\} = \int_0^\infty (1 - \sum_{l\in E} Q(k,l,y))dy$.

This example illustrates how the transient and steady-state distributions of a semi-regenerative process can be characterized through Markov renewal theory. This is especially important in queueing theory since the number of customers in the system in most M/G or G/M type non-Markov queues possess semi-regenerative properties, as illustrated in Sections 3.2 and 3.3.

7. REFERENCES

1. Bhat,U.N. 1984. *Elements of Applied Stochastic Processes.* 2nd Edition. Wiley, New York.

2. Çınlar,E. 1975. *Introduction to Stochastic Processes.* Prentice-Hall, Englewood Cliffs, New Jersey.

3. Kendall,D.G. 1953. Stochastic Processes Occuring in the Theory of Queues and Their Analysis by the Method of the Embedded Markov Chain. *Ann. Math. Stat.* **34**, 338-354.

4. Ross,S.M. 1983. *Stochastic Processes.* Wiley, New York.

Chapter 2

ELEMENTARY QUEUEING MODELS

Walid Moussattat
Higher Institude of
Applied Science and Technology
Damascus, Syria

1. INTRODUCTION

Queueing theory was initiated by the Danish engineer Erlang who studied it in relation with telecommunication problems. Later, it was found that the same concepts had important applications in operations research in the context of service systems. From the beginning of the sixties, interest in queueing theory was further stimulated by computer applications especially in computer networks. During all this time the mathematical theory was gradually elucidated, simplified, unified and tied up with the more general theory of stochastic processes, so that queueing theory now represents one of the most important fields of applications of stochastic processes.

A queueing problem arises every time we have customers demanding some services which may not always be immediately available, and customers are forced to queue for the required services. We always assume that randomness intervenes as an ingredient of queueing problems, so that we assume the necessity of probabilistic concepts to model them.

In analyzing and modeling queueing problems, the following three notions seem to emerge naturally:

1. The arrival of customers,
2. The service duration,
3. The service strategy or policy,

and queues are studied and classified according to the assumptions introduced on the above notions.

Usually we assume that the flow of customers form a stochastic process, or more precisely a point process $A = \{A_t; t \geq 0\}$ where for every t, A_t represents a random variable which is the number of customers who arrive to demand services during $[0, t]$. The process A is specified in queueing theory in one of the following two ways:

1. In elementary models we assume that A is a Poisson process,

2. We postulate in some cases that the probability distribution between two consecutive arrivals is of a special form (for example, exponential or any general distribution).

Of course, it is possible to deduce the Poisson structure of the process or the probabilities from plausible properties which we can impose on the events or the random variables under consideration. Similar assumptions are usually introduced concerning service durations which are, in elementary models, assumed to be exponentially distributed and independent of the flow of customers.

By service strategy or policy we mean the restriction imposed on the queue and the way a customer is chosen from the queue for servicing. The service policy and the remaining assumptions are used to construct the mathematical model for the queue under study. This model can be expressed in several forms:

1. In elementary queueing models, it is a system of ordinary differential equations whose solution gives the probability that the system is at a given state at a given time t,
2. In most cases, the solutions of such systems of differential equations are difficult, and we seek only what is called the steady-state solution, i.e., the solution to which the systems will stabilize in the long run.

Usually 1 and 2 allow us to obtain most of the probabilities of interest in practice for elementary problems, but in many cases we are also interested in the exit process from a given queue.

Once a model and its consequences (either qualitative or quantitative) are elaborated, we proceed as follows:

1. Simulate real queues and examine whether the results deduced theoretically are confirmed in reality,
2. If it is seen that some model is capable of explaining situations where the basic assumptions are different from those of the model, a problem of interest would be to try to prove that the same conclusion hold under different assumptions. This is the case in some queueing problems in service networks.

Of course, when all this has been accomplished in theory for certain fields, we try either to enlarge its scope by considering other applications in other fields or to construct new models in the same theory for newer problems.

2. POISSON ARRIVAL PROCESS AND RELATED RESULTS

2.1. The Poisson Process

The most widely used cusromer arrival process in many queueing models is the Poisson process which obeys the following properties:

1. The number of customers to arrive in two disjoint time intervals are independent; this is expressed by saying that the arrival process A has independent increments. Hence, $A_{t+s} - A_t$ is independent of $\{A_u; \ u \leq t\}$ for all $t, s \geq 0$. Moreover, A also has stationary increments, i.e., $A_{t+s} - A_t$ has the same distribution as A_s for all $t, s \geq 0$.
2. The probability of at least one arrival occuring in a time period of duration h is equivalent to λt when h is small, i.e.,

$$P\{A_h \geq 1\} = \lambda h + o(h)$$

   for some $\lambda \geq 0$.
3. The probability of two or more arrivals occuring during time h is $o(h)$ so that $P\{A_h \geq 2\} = o(h)$.

We suppose $A_0 = 0$, and using the distribution given above we can deduce that for all $s \geq 0$

$$P\{A_s = k\} = \frac{e^{-\lambda t}(\lambda t)^k}{k!} \qquad k = 0, 1, \ldots$$

so that A_t has the Poisson distribution with parameter λt. This implies that $E[A_t] = \lambda t$ and $\text{Var}(A_t) = \lambda t$, so that λ represents $E[A_1]$, i.e., the expected value or average of the number of customers that will arrive in unit time. Since $t \to A_t$ is an integer valued function, the trajectory of Poisson processes is a step function with jumps ocurring at arrival of customers.

2.2. Arrival Time Process of a Poisson Process

Let $T_1 = \inf\{t \geq 0; A_t = 1\}$ be the time of occurence of the first arrival, $T_2 = \inf\{t \geq 0; A_t = 2\}$ be the time of occurence of the second arrival, and in general, $T_n = \inf\{t \geq 0 : A_t = n\}$ be the time of occurence of the nth arrival with $T_0 = 0$. The interarrival times are defined by $W_1 = T_1 - T_0$, $W_2 = T_2 - T_1$, and in general, $W_n = T_n - T_{n-1}$.

Note that $A_t \leq n \Leftrightarrow T_{n+1} > t$ and that $A_t = n \Leftrightarrow (T_n \leq t \text{ and } T_{n+1} > t)$, because $A_t \leq n$ implies that up to the time t, at most n arrivals have occured; hence, the time of arrival of customer $n + 1$ is strictly after t. Therefore, $P\{A_t \leq n\} = P\{T_{n+1} > t\}$ and

$$P\{T_{n+1} \leq t\} = 1 - P\{A_t \leq n\} = 1 - \sum_{k=0}^{n} \frac{e^{-\lambda t}(\lambda t)^k}{k!} \ .$$

Similarly, $P\{A_t = n\} = P\{T_n \leq t\} - P\{T_{n+1} \leq t\}$ and, hence, $P\{A_t = 0\} = 1 - P\{T_1 \leq t\}$. Thus,

$$P\{W_1 > t\} = P\{T_1 > t\} = 1 - P\{T_1 \leq t\} = P\{A_t = 0\} = e^{-\lambda t}$$

and we observe that W_1 has an exponential distribution with parameter λ. This implies that

$$P\{W_1 \leq t\} = 1 - e^{-\lambda t} \Rightarrow d(P\{W_1 \leq t\})/dt = \lambda e^{-\lambda t} , \qquad t \geq 0$$

and $E[W_1] = 1/\lambda$ and $\text{Var}(W_1) = 1/\lambda^2$. As a matter of fact, it is possible to show that the interarrival times between successive arrivals in a homogeneous Poisson processes with rate λ are independent and identically distributed with the exponential distribution with mean $1/\lambda$, so that W_n has density $\lambda e^{-\lambda x}$, for all $n \geq 1$. For an elementary, but rather long proof the reader can refer to Hoel et al. (1971, p.231). It is possible to prove that this characterizes the Poisson process, i.e., if we start with independent and identically distributed random variables $\{W_n\}$ whose distributions are all exponential with mean $1/\lambda$, and define the sequence of sums $T_1 = W_1$, $T_2 = W_1 + W_2$, and in general, $T_n = W_1 + W_2 + \cdots + W_n$, and associate with this sum the counting process defined by $A_t = \sup\{n \geq 0 : T_n \leq t\}$, then A is a Poisson process with rate λ. A proof of this fundamental result presented with the terminology of renewal processes can be found in Parzen (1964, p.174).

These results show that an arrival process is an ordinary Poisson process with rate λ if and only if the interarrival times are independent and identically distributed with the exponential distribution with mean $1/\lambda$. A typical realization of a Poisson process is given in Figure 1 which illustrates the relationship between the Poisson process A, the arrival time process $\{T_n\}$, and the interarrival time process $\{W_n\}$.

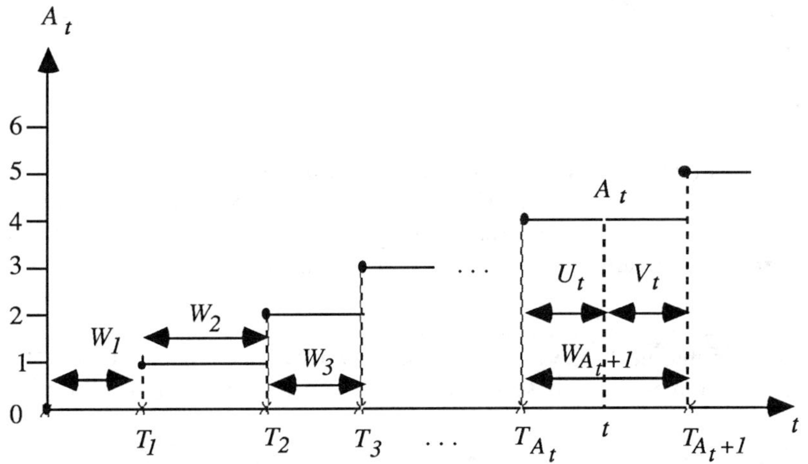

Figure 1. A Typical Realization of a Poisson Process

To establish the Markov character of some queueing processes, we need to understand the remanining waiting time, or forward recursive time, at t defined by $V_t = T_{A_t+1} - t$. This means that at time t, A_t arrivals have occured and the next arrival will occur at time T_{A_t+1} so that we should wait for $V_t = T_{A_t+1} - t$ until the next arrival. This is also called excess life in the terminology of renewal processes. Note that $V_t > x$

if and only if there are no arrivals in the interval $(t, t + x]$ so that

$$P\{V_t > x\} = P\{A_{t+x} - A_t = 0\}$$
$$= P\{A_x = 0\} = e^{-\lambda x} , \qquad x > 0 .$$

Hence, for a Poisson process the remaining waiting time is also exponentially distributed (like the interarrival time) with the same mean $1/\lambda$. The backward recurrence time, or the time since the last arrival, at time t is $U_t = t - T_{A_t}$ as shown in Figure 1. Note that $U_t \leq t$, and $U_t > x \Leftrightarrow A_t - A_{t-x} = 0$ for $x < t$, which occurs with probability $e^{-\lambda x}$. Hence,

$$P\{U_t \leq x\} = \begin{cases} 1 - e^{-\lambda x} & ,0 \leq x < t \\ 1 & ,x \geq t \end{cases}$$

We can determine the joint distribution of V_t and U_t since for $0 < y < t$ we have

$$(V_t > x, \ U_t > y) \Leftrightarrow A_{t+x} - A_{t-y} = 0$$

which occurs with probability $e^{-\lambda(x+y)}$, so that

$$P\{V_t > x, \ U_t > y\} = \begin{cases} e^{-\lambda(x+y)} & ,\text{if } 0 < y < t \\ 0 & ,\text{if } y \geq t \end{cases}$$

Another result which is useful for simulation is that the conditional distribution of $T_1, T_2, ..., T_n$ given $\{A_t = n\}$ is the same as the distribution of order statistics constructed with n independent uniformly distributed random variables on $[0, t]$. In other words,

$$P\{T_i \leq t_i, \ i = 1, 2, ..., n | A_t = n\} = \frac{n!}{t^n} \int_0^{t_1} \cdots \int_{x_{n-2}}^{t_{n-1}} \int_{x_{n-1}}^{t_n} dx_1 \ldots dx_n$$

and a proof can be found in Karlin and Taylor (1975, p.126).

2.3. Exponential and Erlang Distributions

What makes the exponential distribution important in queueing theory is its memorylessness property. As a matter of fact the exponential distribution is the only distribution without memory. In other words, X is a random variable satisfying the following memorylessness property

$$P\{X > x + y | X > x\} = P\{X > y\} , \qquad x, y \geq 0$$

if and only if it has the exponential distribution

$$P\{X > x\} = e^{-\lambda x}$$

for some $\lambda \geq 0$. This property is extremely important in queueing systems because if, for example, interarrival times are exponentially distributed, then the time until the next arrival is exponentially distributed independent of how much time has passed since the last arrival. Similarly, if service durations are exponentially distributed then the time until service completion is still exponentially distributed independent of how much time has passed since the start of the service. Thus, no information from the past is required to make future predictions on these durations. This generally leads to Markov models which can be analyzed using available results on Markov processes.

Erlang distribution appears in cases where services are performed in stages so that it is the sum of independent and identically distributed random variable which have an exponential distribution, or when arrivals occur in batches with customers requiring exponentially distributed services. It was Erlang who first recognized that most of the basic computations of queueing theory can be carried out when we replace the exponential distribution by the Erlang distribution.

Since the Laplace transform of the exponential distribution with parameter λ is

$$E[e^{-\alpha X}] = \frac{\lambda}{\alpha + \lambda} \,,$$

the Laplace transform of the sum $X_1 + X_2 + \cdots + X_n$ of n independent exponentially distributed random variables with parameter λ is

$$E[e^{-\alpha(X_1 + \cdots + X_n)}] = \left(\frac{\lambda}{\alpha + \lambda}\right)^n \,.$$

This is known as the Erlang (n) distribution which has the density

$$\frac{dP\{X_1 + X_2 + \cdots + X_n \leq x\}}{dx} = \lambda \frac{(\lambda x)^{n-1}}{(n-1)!} e^{-\lambda x} \,, \qquad x \geq 0 \,.$$

It also follows from the above explanation that the time of the nth arrival T_n of a Poisson process has the Erlang(n) distribution.

3.　CLASSIFICATION OF QUEUEING MODELS

The classification of queues is done according to the assumptions on the customer arrival process, the distribution of service durations, and the service policy.

3.1. Arrival Process

The above results show that it is natural in elementary queueing theory to start with the assumption that the flow of demands is an ordinary Poisson process with rate λ, or equivalently, the interarrival times are independent and identically distributed with the exponential function with parameter λ. This arrival process is denoted by the letter M. Similarly, we denote by G the case where interarrival times between customers are independent and identically distributed with some common general distribution.

3.2. Service Durations

Due to the memoryless property of the exponential distribution, which is desirable in many cases, it is assumed in elementary queueing theory that service durations are exponentially distributed with mean $1/\mu$. Usually we assume that services are independent of customer arrivals, though there should be a relation between λ and μ if we want the size of the queue not to increase without limit. We will also denote this assumption by M, and by G, the more general case of an arbitrary distribution.

3.3. Service Policy

In this regard, the following three items must be specified:

1. The number of servers,
2. The capacity of the waiting room where those customers waiting in the queue can be kept,
3. The way we choose a customer from the queue, or the service discipline.

If we assume that there are m servers, $1 \leq m \leq \infty$, then all of the first m customers can be served immediately. The customers arriving when all the m servers are busy are either lost, if the waiting room capacity is $k = 0$ (we say that we have a system with refusal when $k = 0$), or if the queue capacity is $k > 0$, we refuse those customers beyond the queue capacity k. Of course, if the queue capacity is $k = \infty$ then all arriving customers are allowed to queue. We assume in these cases that all customers who are admitted for queueing necessarily wait until they are served.

In some cases it is assumed that customers wait a random amount of time, and if they are still unserved by the end of this time they depart without being served. These customers are said to renege from the queue. Sometimes, customers may refuse to join a queue if it is too long. This is referred to as balking.

There exists several rules to choose a customer out of an existing queue for service, for example:

1. The rule first in, first out (FIFO) is usually used in many cases where service is rendered to humans,
2. The rule last in, first out (LIFO) may be more useful in queueing models of manufacturing systems,
3. The rule may involve servicing in random order (SIRO) where customers are selected randomly from the queue,
4. Priority rules (PR) generally apply in cases where customers are classified according to their service requirements, as in emergency rooms of hospitals.

It was Kendall (1953) who initially proposed the following notation used to summarize the assumptions in classifying queues:

$$\text{A/B/C/k (service rule)}$$

where

A = abbreviation for the interarrival distribution,

B = abbreviation for the service time distribution,

C = number of servers,

k = queue capacity,

and the service rule specifies the way customers are selected from the queue for servicing. Actually, Kendall's notation included only the first three elements (i.e., A, B, and C), and this was later extended to the form given above by Lee (1966).

The following are some examples:

1. $M/M/m/k$ (FIFO) will denote a queue where interarrival times are exponential with mean $1/\lambda$, service times are exponential with mean $1/\mu$, there are m servers, queue capacity is k, and the first-in-first-out rule is used in selecting customers for service. If $k = \infty$, then the queue $M/M/m/\infty$ (FIFO) is usually denoted by $M/M/m$ (FIFO), and further simplified to $M/M/m$ when the rule FIFO is understood as default.
2. $G/M/m$ will denote a queue with general type distribution for interarrival times, exponential service times, m servers, infinite queue capacity, and FIFO rule for selecting customers.
3. The queue $M/G/m$ is defined similarly, and the most general type is denoted by $G/G/m/k$.

4.　TRANSIENT ANALYSIS FOR SOME M/M QUEUES

In this section, we will derive for some M/M queues, systems of differential equations, or the so-called difference-differential equations, governing their states. In all of the models, we assume that the arrival process is Poisson with rate λ, and service durations are exponentially distributed with rate μ.

4.1.　$M/M/m$ Queues with Refusal and FIFO Service Discipline

In this model, we have m servers and the system can be in one of the $(m+1)$ states in $\{0, 1, 2, ..., m\}$ where state i implies that there are i customers in the system. Note that arriving customers do not joint the queue, or they are refused, when the system is full. This implies that the queue capacity is $k = 0$.

Let $P_k(t)$ denote the probability that the system is in state k at time t. Consider the case where $k = 0$, then the system will be in state 0 at time $t + h$ if one of the following mutually exclusive events takes place:

1. The system was in state 0 at time t and remains in it in the interval $(t, t+h)$. Hence no customer arrives, and the probability of this is

$$P_0(t)[1 - \lambda h + o(h)] ,$$

2. The system was in state 1 at time t, no demand occurs during $(t, t + h)$ and the server finishes the service in $(t, t + h)$. The probability of this is

$$\begin{aligned} P_1(t)[1 - \lambda h + o(h)][1 - e^{-\mu h}] &= P_1(t)[1 - \lambda h + o(h)][\mu h + o(h)] \\ &= P_1(t)[\mu h + o(h)] , \end{aligned}$$

3. All of the other possible transitions to 0 have $o(h)$ probabilities.

Combining the terms in cases 1,2, and 3, we obtain

$$P_0(t + h) = P_0(t)[1 - \lambda h + o(h)] + P_1(t)[\mu h + o(h)] + o(h)$$

Therefore,

$$\frac{dP_0(t)}{dt} = -\lambda P_0(t) + \mu P_1(t) .$$

Consider next the case where $0 < k < m$. The system will be in state k at time $t + h$ if one of the following mutually exclusive events takes place:

1. At time t, the system was in state k and in the interval $(t, t+h)$, neither a customer arrives nor a customer departs. The probability of this is

$$\begin{aligned} P_k(t)[1 - \lambda h + o(h)][e^{-\mu h}]^k &= P_k(t)[1 - \lambda h + o(h)][1 - \mu k h + o(h)] \\ &= P_k(t)[1 - (\lambda + \mu k)h + o(h)] , \end{aligned}$$

2. At time t, the system was in state $k - 1$, a customer arrives in the interval $(t, t+h)$, but no service is finished. The probability of this is

$$\begin{aligned} P_{k-1}(t)[\lambda h + o(h)][e^{-\mu h}]^{k-1} &= P_{k-1}(t)[\lambda h + o(h)][1 - \mu h]^{k-1} \\ &= P_{k-1}(t)[\lambda h + o(h)] , \end{aligned}$$

3. At time t, the system was in state $k + 1$, no customer arrives but one out of the $(k + 1)$ services is finished during $(t, t + h)$. The probability of this is

$$\begin{aligned} P_{k+1}(t)[1 - \lambda h + o(h)][1 - (e^{-\mu h})^{k+1}] &= P_{k+1}(t)[1 - \lambda h + o(h)][(k + 1)\mu h + o(h)] \\ &= P_{k+1}(t)[(k + 1)\mu h + o(h)] , \end{aligned}$$

4. All of the other possible transitions to k have $o(h)$ probability.

Combining the corresponding terms in 1,2,3, and 4, we obtain

$$P_k(t+h) = P_k(t)[1 - (\lambda + \mu k h + o(h)] + P_{k-1}[\lambda h + o(h)]$$
$$+ P_{k+1}[(k+1)\mu h + o(h)] + o(h)$$

Therefore,

$$\frac{dP_k(t)}{dt} = \lambda P_{k-1}(t) - (\lambda + k\mu)P_k(t) + (k+1)\mu P_{k+1}(t) \ .$$

A similar argument holds for $k = m$ and yields

$$\frac{dP_m(t)}{dt} = -\lambda P_{m-1}(t) + \mu P_m(t) \ .$$

These equations for $0 \leq k \leq m$ are called difference-differential equations of queueing theory, and defining the vector functions

$$P(t) = [P_0(t), P_1(t), ..., P_m(t)]$$

and

$$\frac{dP(t)}{dt} = \left[\frac{dP_0(t)}{dt}, \frac{dP_1(t)}{dt}, ..., \frac{dP_m(t)}{dt}\right]$$

this system of equations can be written in the form

$$\frac{dP(t)}{dt} = P(t)A$$

where A is the matrix

$$A = \begin{pmatrix}
-\lambda & \lambda & 0 & . & . & . & . \\
\mu & -(\lambda + \mu) & \lambda & 0 & . & . & . \\
0 & 2\mu & -(\lambda + 2\mu) & \lambda & . & . & . \\
. & . & . & . & . & . & . \\
. & . & . & . & . & . & . \\
. & . & . & . & . & . & . \\
. & . & . & . & 0 & m\mu & -m\mu
\end{pmatrix}$$

and the boundary condition is $P_0(0) = 1$, $P_k(0) = 0$ for $1 \leq k \leq m$.

4.2. $M/M/m/k$ Queue

We now assume that there are m servers, the queue capacity is $k > 0$, and customers who arrive when the queue is full do not join the queue. There are $m + k + 1$ states for the system given by $\{0, 1, ..., m + k\}$. It is clear that $P_0(t), ..., P_{m-1}(t)$ satisfy the same equations established before. As for $P_m(t)$, we have

$$\frac{dP_m(t)}{dt} = \lambda P_{m-1}(t) - (\lambda + m\mu)P_m(t) + m\mu P_{m+1}(t)$$

Similarly, we obtain

$$\frac{dP(t)_{m+s}}{dt} = \lambda P_{m+s-1}(t) - (\lambda + m\mu)P_{m+s}(t) + m\mu P_{m+s+1}(t)$$

for $1 \le s < k$ and

$$\frac{dP_{m+k}(t)}{dt} = \lambda P_{m+k-1}(t) - m\mu P_{m+k}(t)$$

which we can put, once again, in matrix form as $dP(t)/dt = P(t)A$ where

$$A = \begin{pmatrix} -\lambda & \lambda & 0 & . & . & & . & & . & & . \\ \mu & -(\lambda + \mu) & \lambda & 0 & . & & . & & . & & . \\ 0 & 2\mu & -(\lambda + 2\mu) & \lambda & . & & . & & . & & . \\ . & . & . & . & . & & . & & . & & . \\ . & . & . & . & . & & . & & . & & . \\ . & . & . & . & . & m\mu & -(\lambda + m\mu) & & \lambda & & . \\ . & . & . & . & . & & . & & . & & . \\ . & . & . & . & . & & . & & . & & . \\ . & . & . & . & . & & m\mu & -(\lambda + m\mu) & & \lambda \\ . & . & . & . & . & & 0 & m\mu & & -m\mu \end{pmatrix}$$

with the boundary condition $P_0(0) = 1$, $P_j(0) = 0$ for $j = 1, 2, ..., m + k$.

4.3. $M/M/m$ Queues with Exponential Waiting Time

Suppose that a customer who finds all servers busy waits for a random amount of time T and departs without being served if his service does not start until then. In this case, the system has an infinite number of states, and since $k = \infty$. Assuming T has an exponential distribution with $E[T] = 1/\nu$, independent of services and arrivals, we can obtain the following system of difference-differential equations:

$$\frac{dP_0(t)}{dt} = -\lambda P_0(t) + \mu P_1(t)$$

$$\frac{dP_k(t)}{dt} = \lambda P_{k-1}(t) - (\lambda + k\mu)P_k(t) + (k + 1)\mu P_{k+1}(t) , \qquad 1 \le k \le m - 1$$

$$\frac{dP_m(t)}{dt} = \lambda P_{m-1}(t) - (\lambda + m\mu)P_m(t) + (m\mu + \nu)P_{m+1}(t)$$

$$\frac{dP_{m+s}(t)}{dt} = \lambda P_{m+s-1}(t) - (\lambda + m\mu + s\nu)P_{m+s}(t) + [m\mu + (s + 1)\nu]P_{m+s+1}(t)$$

for $s \geq 1$. This satisfies $dP(t)/dt = P(t)A$ where

$$A = \begin{pmatrix}
-\lambda & \lambda & 0 & & . & & . & & . \\
\mu & -(\lambda+\mu) & \lambda & 0 & & . & & . \\
0 & 2\mu & -(\lambda+2\mu) & \lambda & & . & & . \\
. & . & . & . & & . & & . \\
. & . & . & . & & . & & . \\
. & m\mu & -(\lambda+m\mu) & \lambda & & . & & . \\
. & . & m\mu+\nu & -(\lambda+m\mu+\nu) & \lambda & & . \\
. & . & . & m\mu+2\nu & -(\lambda+m\mu+2\nu) & \lambda \\
. & . & . & . & & . & & .
\end{pmatrix}$$

with the initial conditional $P_0(0) = 1$ and $P_j(0) = 0$ for all $j \geq 1$. We must also add the condition that $\sum P_k(t) = 1$ since, without it, it is possible to get a solution such that $\sum P_k(t_0) < 1$ for some $t_0 \geq 0$.

4.4. The Solution of the Difference-Differential Equations

To find a solution satisfying the difference-differential equations, $dP(t)/dt = P(t)A$, one method is to use operator theory. For the finite state case, the solution is

$$P(t) = P(0)e^{tA}$$

where

$$e^{tA} = \sum_{n=0}^{\infty} \frac{t^n}{n!}A^n$$

For the infinite state cases, however ,the operator A is usually unlounded, and we need Hill-Yosida semi-group theory.

Another method preferred by engineers to solve these equations is to introduce the generating function

$$G(z,t) = \sum_{k=0}^{\infty} P_k(t)z^k , \qquad z,t \geq 0$$

which satisfies

$$\frac{\partial G(z,t)}{\partial t} = \sum_{k=0}^{\infty} \frac{dP_k(t)}{dt}z^k , \qquad z,t \geq 0 .$$

By substituting $dP_k(t)/dt$ for their values as a function of $\{P_k(t)\}$ we obtain a partial differential equation which we may try to solve using Laplace transforms. Once $G(z,t)$ has been obtained we may try to expand it as a power series in z to obtain $\{P_k(t)\}$ as the correponding coefficients of $\{z^k\}$ in the power series. Except for a few particular cases, this approach is difficult and not very useful, because we are interested not only on the solution as a function of t, but more on where this solution stabilizes. In other words, we are not only interested in the transient solution but also in the steady-state solution $\lim_{t\to\infty} P_k(t)$, $k = 0, 1, \dots$.

4.5. M/M Queues and Markov Processes

Let N_t denote the number of customers in the system, including those being served, at time t. Then, in M/M queues N is a continuous time Markov chain or the Markov process of Section 4 in Chapter 2. Since the customer arrival process is Poisson at any time t, the distribution of the waiting time until the next arrival is exponential with mean $1/\lambda$, independent of the time elapsed since the last arrival. Also, the remaining service times have the exponential distribution with mean $1/\mu$ independent of how long they have been going on due to the memorylessness property. These imply that the future evolution of the queue after any time t depends only on the present state N_t, and not on its past. Thus, the Markov property is satisfied. Note that

$$P\{N_t = j\} = P\{N_t = j | N_0 = 0\} = P_t(0, j)$$

since we assume that the system is empty at time 0, i.e., $N_0 = 0$. Thus, what we denoted by $P_j(t)$ is the preceeding subsection is actually $P_{0j}(t)$ with the present notation. The difference-differential equations can now be written as

$$\left[\frac{P_t(0,0)}{dt}, \dots, \frac{P_t(0,j)}{dt}, \dots \right] = [P_t(0,0), \dots, P_t(0,j), \dots] A$$

or $P_t = P_t A$ in compact notation where A is the generator of the continuous-time Markov chain N defined as $A(i, j) = \lim_{h\to 0}(P_h(i, j) - P_0(i, j))/h$.

Assuming that $P_0 = I$, as is trivially the case for N in queueing models, we can deduce that for any Markov process

$$\frac{P_{t+h} - P_t}{h} = \frac{P_t P_h - P_t}{h} = P_t \frac{P_h - I}{h}$$
$$= \frac{P_h P_t - P_t}{h} = \frac{P_h - I}{h} P_t$$

using the Chapman-Kolmogorov forward and backward equations ,i.e., $P_{t+h} = P_t P_h = P_h P_t$. This shows that if $\lim_{h\to 0}(P_h - I)/h = A$ exists, then $dP_t/dt = AP_t$ and $dP_t/dt = P_t A$ which are known as the Chapman-Kolmogorov backward and forward differential equations respectively. Thus, the difference-differential equations obtained in Sections 4.1, 4.2, and 4.3 are nothing else but the Chapman-Kolmogorov differential equations. Moreover, these M/M models are particular cases of the so-called birth and death processes which are discussed in some detail in Section 4.6 of Chapter 1. These are continuous-time homogeneous Markov chains such that $P_h(i, i+1) = \lambda_i h + o(h)$, i.e., the probability of birth is equivalent to $\lambda_i h$ when h is small; and $P_h(i, i-1) = \mu_i h + o(h)$, i.e., the probability of death is equivalent to $\mu_i h$ when h is small: Moreover,

$$P_h(i, i) = 1 - (\lambda_i + \mu_i)h + o(h)$$

with the boundary condition $P_0(i, j) = I(i, j)$. In birth and death models, λ_i and μ_i represent the customer arrival rate and service or departure rate when there are i customers in the system respectively.

To analyze the limiting or steady-state distribution of the Markov process N with state space E, let $p_k(s) = P\{N_s = k\}$, then

$$\begin{aligned}
p_k(s+t) &= P\{N_{s+t} = k\} \\
&= \sum_{j \in E} P\{N_{s+t} = k | N_s = j\} P\{N_s = j\} \\
&= \sum_{j \in E} p_j(s) P_t(j, k)
\end{aligned}$$

or $p(s+t) = p(s)P_t$ in compact form where $p(t)$ is the vector function consisting of $\{p_k(t)\}$. Note that this gives

$$\frac{p(s+t) - p(s)}{t} = p(s)\frac{P(t) - I}{t}$$

so that $dp(s)/ds = p(s)A$ where A is the generator of N.

If μ is a vector satisfying $\mu A = 0$ with $\mu 1 = 1$, then the initial condition $p(0) = \mu$ implies that $p(s) = \mu$ will satisfy the equation $dp(s)/ds = p(s)A$ since $dp(s)/ds = 0 = p(s)A = \mu A$. By the uniqueness of the solution, we see that μ will also satisfy $\mu = \mu P_t$ for all $t \geq 0$. The distribution μ is called the stationary distribution of the Markov process N since if $P\{N_0 = k\} = \mu(k)$, then $P\{N_t = k\} = \sum_i \mu(i) P_t(i, k) = \mu P_t(k) = \mu(k)$ independent of time. This is also called the limiting or steady-state distribution of N since $\lim_{t \to \infty} P\{N_t = k\} = \mu(k)$ trivially.

If we denote

$$A = \begin{pmatrix}
-q_1 & q_{12} & q_{13} & \cdot & \cdot & \cdot \\
q_{21} & -q_2 & q_{23} & \cdot & \cdot & \cdot \\
q_{31} & q_{32} & -q_3 & \cdot & \cdot & \cdot \\
\cdot & \cdot & \cdot & \cdot & \cdot & \cdot \\
\cdot & \cdot & \cdot & \cdot & \cdot & \cdot
\end{pmatrix}$$

so that $q_j = \lim_{t \to 0}(1 - P_t(j, j))/t$, $q_{jk} = \lim_{t \to 0}(1 - P_t(j, k))/t$ with $q_j \geq 0$, $q_{jk} \geq 0$, then $\mu A = 0$ can be written as

$$\mu(j)q_j = \sum_{k \neq j} \mu(k)q_{kj} \, ,$$

but since $\sum_k P_t(j,k) = 1$ we have

$$\frac{1 - P_t(j,j)}{t} = \sum_{k \neq j} \frac{P_t(j,k)}{t}$$

and thus

$$q_j = \sum_{k \neq j} q_{jk}$$

so that

$$\mu(j)q_j = \sum_{k \neq j} \mu(k)q_{kj} = \sum_{k \neq j} \mu(j)q_{jk} \ .$$

If we call q_{jk} the flow per unit time from state j to state k, then the last equation expresses the equality of two flows, $\sum_{k \neq j} \mu(k)q_{jk}$, which is the flow per unit time which enter into state j, and $\sum_{k \neq j} \mu(j)q_{jk}$, which is the flow per unit time which goes out of state j. This equality is called the global balance equation for state j. The following subsection illustrates how the steady-state distributions can be found for the M/M models in Section 4.1, 4.2, and 4.3. We shall, however, make a slight change in our notation so that the steady-state distribution is now denoted by

$$p_k = \lim_{t \to \infty} P\{N_t = k\} \ , \qquad k = 0, 1, \dots \ ,$$

instead of $\{\mu_k\}$, which is the notation reserved for service rates.

5. STEADY-STATE ANALYSIS FOR SOME M/M QUEUES

5.1. $M/M/m$ Queues with Refusal and FIFO Service Discipline

It is clear from Section 4.1 that the system of equations to obtain the steady-state solution $\{p_k\}$ is given by $pA = 0$, or

$$-\lambda p_0 + \mu p_1 = 0$$
$$\lambda p_{k-1} - (\lambda + k\mu)p_k + (k+1)\mu p_{k+1} = 0 \ , \qquad 1 \leq k \leq m-1 \ ,$$
$$\lambda p_{m-1} - m\mu p_m = 0$$

with $\sum_{k=0}^{m} p_k = 1$. The first equation gives $p_1 = (\lambda/\mu)p_0$, thus

$$\begin{aligned}
p_2 &= (1/2\mu)[-\lambda p_0 + (\lambda + \mu)p_1] \\
&= (1/2\mu)[-\lambda p_0 + (\lambda + \mu)(\lambda/\mu)p_0] \\
&= p_0 \lambda^2 / 2\mu^2
\end{aligned}$$

and more generally,

$$p_k = p_0 \frac{\lambda^k}{k! \mu^k} .$$

If we introduce

$$\rho = \frac{\lambda}{\mu} = \frac{1/\mu}{1/\lambda} = \frac{\text{Expected service duration}}{\text{Expected interarrival time}}$$

then

$$p_k = \frac{\rho^k}{k!} p_0 .$$

Hence

$$\sum_{k=0}^{m} p_k = p_0 \sum_{k=0}^{m} \frac{\rho^k}{k!} = 1$$

gives

$$p_0 = \frac{1}{\sum_{k=0}^{m} \rho^k / k!}$$

which is the probability that all servers are idle. Therefore,

$$p_k = (\rho^k / k!) \Big/ \sum_{j=0}^{m} (\rho^j / j!) .$$

In particular, the probability of refusing a demand is $P_{ref} = p_m$. For the one server case where $m = 1$, $P_{ref} = p_1 = \rho/(1+\rho)$ and the probability of the system being idle is $P_0 = 1 - P_{ref} = 1/(1+\rho)$.

5.2. $M/M/m/k$ Queues

Using the generator A given in Section 4.2 we now have the following system of difference equations

$$-\lambda p_0 + \mu p_1 = 0$$
$$\lambda p_{j-1} - (\lambda + j\mu)p_j + (j+1)\mu p_{j+1} = 0 , \quad 1 \le j \le m - 1$$
$$\lambda p_{m-1} - (\lambda + m\mu)p_m + m\mu p_{m+1} = 0$$
$$\lambda p_{m+s-1} - (\lambda + m\mu)p_{m+s} + m\mu p_{m+s+1} = 0 , \quad 1 \le s \le k$$
$$\lambda p_{m+k+1} - m\mu p_{m+k} = 0$$

with $\sum_{j=0}^{m+k} p_j = 1$. The solution of this system of equations can be obtained to be

$$p_j = \frac{\rho^j / j!}{\sum_{k=0}^{m} \rho^k / k! + (\rho^m / m!) \sum_{i=1}^{k} (\rho/m)^i} , \quad 0 \le j \le m$$

and

$$p_{m+s} = \frac{\left(\rho^m/m!\right)\left(\rho/m\right)^s}{\sum_{k=0}^{m}\rho^k/k! + \left(\rho^m/m!\right)\sum_{i=1}^{k}\left(\rho/m\right)^i} \,, \qquad 1 \le s \le k$$

where $\rho = \lambda/\mu$. The probability of refusal, i.e., the probability of a customer leaving the system without being served is p_{m+k}.

5.3. $M/M/m$ Queue with Exponential Waiting Time

It follows from Section 4.3 that the steady-state equations now become

$$-\lambda p_0 + \mu p_1 = 0$$
$$\lambda p_{k-1} - (\lambda + k\mu)p_k + (k+1)\mu p_{k+1} = 0 \,, \qquad 1 \le k \le m-1$$
$$\lambda p_{m-1} - (\lambda + m\mu)p_m + (m\mu + \nu)p_{m+1} = 0$$
$$\lambda p_{m+s-1} - (\lambda + m\mu + s\nu)p_{m+s} + [m\mu + (s+1)\nu]p_{m+s+1} = 0 \,, \qquad s \ge 1$$

with $\sum_{k=0}^{\infty} p_k = 1$. We can solve these equations and find that

$$p_k = \frac{\rho^k}{k!}p_0 \,, \qquad 0 \le k \le m$$

$$p_{m+s} = \frac{\rho^{m+s} p_0}{m! \prod_{n=1}^{s}(m + n\beta)} \,, \qquad s \ge 1$$

where

$$p_0 = \frac{1}{\sum_{k=0}^{m}\rho^k/k! + \left(\rho^m/m!\right)\sum_{s=1}^{\infty}\left(\rho^s \Big/ \prod_{n=1}^{s}(m+n\beta)\right)}$$

and $\rho = \lambda/\mu$, $\beta = \nu/\mu$.

Note that as $\beta \to \infty$ this system becomes an $M/M/m$ queue with refusal and

$$p_k = \begin{cases} \dfrac{\rho^k/k!}{\sum_{j=0}^{m}\rho^j/j!} & ,0 \le k \le m \\ 0 & ,k \ge m+1 \end{cases}$$

On the other hand, if $\beta \to 0$ so that customers do not leave without being served, then

$$p_0 = \frac{1}{\sum_{k=0}^{m}\rho^k/k! + \left(\rho^m/m!\right)\sum_{s=1}^{\infty}\rho^s/m^s} \,.$$

Hence, if $\rho/m \ge 1$ then $p_0 = 0$ and so $p_k = 0$ for all $k \ge 1$ so that the queue is unstable and servers will not be able to keep up with services demanded by the arriving customers. This results in a queueing system where the number of customers blows off to infinity. If, on the other hand, $\rho/m < 1$ then $p_0 > 0$ and we have a steady-state distribution given by

$$p_0 = \frac{1}{\sum_{k=0}^{m} \rho^k / k! + \rho^{m+1} / m!(m - \rho)}$$

and

$$p_k = \frac{\rho^k}{k!} p_0 , \qquad 0 \le k \le m ,$$

$$p_{m+s} = \frac{\rho^{m+s}}{m! m^s} p_0 , \qquad s \ge 1 .$$

In this case, we can also compute the expected number of customers waiting for service as

$$\sum_{s=1}^{\infty} s p_{m+s} = \frac{\rho^{m+1}}{(m - 1)!(m - \rho)^2}$$

The results presented above for the case where $\beta = 0$ correspond to the classical $M/M/m$ queueing system with m parallel servers. The special case where $m = 1$ leads to the well-known distribution $p_k = (1 - \rho)\rho^k$, $k = 0, 1, ...$, for the simple $M/M/1$ queue.

6. REFERENCES

1. Cox,D.R. and Smith,W.L. 1961. *Queues.* Methuen, London.

2. Hoel,P.G., Port,S.C., and Stone,S.J. 1971. *Introduction to Probability Theory.* Houghton-Mifflin, Boston, Massachusettes.

3. Karlin,S., and Taylor,H.M. 1975. *A First Course in Stochastic Processes.* Academic Press, New York.

4. Kendall,D.G. 1953. Stochastic Processes in the Theory of Queues and Their Analysis by the Method of Imbedded Markov Chains. *Ann.Math.Stat.* **24**, 338-354.

5. Kleinrock,L¿ 1975. *Queueing Systems: Theory.* Vol I. Wiley, New York.

6. Lee,A.M. 1966. *Applied Queueing Theory.* Macmillan, Toronto.

7. Parzen,E. 1962. *Stochastic Processes.* Holden-Day, San Francisco, California.

Chapter 3

ADVANCED QUEUEING MODELS

Süleyman Özekici
Boğaziçi University
Department of Industrial Engineering
Bebek, Istanbul, Turkey

1. INTRODUCTION

1.1. Overview

This Chapter presents an exposure to advanced queueing models with emphasis on stochastic modeling and analysis, rather than applications. Our aim is actually three-fold: to educate the reader on basic topics, to inform them on the state-of-the-art on stochastic modeling, and to motivate them towards research on current issues in the literature. Due to time limitations, our research oriented presentation will concentrate, in particular, on queueing models with service interruptions where the server interrupts the servicing of a customer due to breakdowns, arrival of higher priority customers, etc.. We shall also emphasize some research on the so-called "server-vacation" models in which the servers take vacations or rest periods of random duration.

The literature on queueing theory and applications date all the way back to early 1900's. A review of the literature covering about a sixty-year sojourn until 1967 can be found in Bhat (1969), and a recent note in Prabhu (1987) presents a classified list of basic textbooks and survey papers written on queueing theory. There are many good textbooks available in libraries, and we have made full use of them in preparing the standard textbook-type material in our presentation. Noteworthy in this context are the works of Cox (1961), Tacáks (1962), Jaiswal (1968), Cohen (1969), Gross and Harris (1974), Kleinrock (1975,1976), Cooper (1981), Heyman and Sobel (1982), and Gelenbe and Pujolle (1987); to cite them in chronological order.

Our notation and terminology will be introduced next. We assume that the reader is previously exposed to M/M type queues with the simple birth-and-death structure. Section 2 covers some advanced Markov models of queueing systems which includes bulk arrivals, batch services, priority servicing, and service interruptions. Semi-Markov queueing models of the M/G and G/M type are the topics of Section 3. Finally,

Section 4 is devoted to other models used basically in G/G type queues. This includes virtual-delay processes and diffusion approximations.

1.2. Notation and Terminology

We now introduce the notation and terminology that will be used throughout this note. Let $N = \{N_t; t \geq 0\}$ be the queue process that represents the number of customers in the system. Under equilibrium conditions, the steady-state of N is denoted by $N_\infty \equiv \lim_{t \to \infty} N_t$. The customer arrival processes is $A = \{A_t \geq 0\}$ with arrival times $\{T_n\}$, and successive customer departure times are represented by the process $\{U_n\}$. Moreover, $\{S_n\}$ represents the service duration of consecutive customers.

We are ultimately interested in the stochastic structure of N. Its probability law,

$$P_n(t) = P\{N_t = n\}\,, \qquad t \geq 0,\ n \geq 0 \tag{1}$$

is perhaps the most sought-after object in queueing theory. This requires the so-called transient analysis which, unfortunately, does not generally yield exact and explicit expressions for $P_n(t)$ in most cases.

On the other hand, the steady-state analysis which primarily seeks for the limiting distribution of N_t, to be denoted by $p_n \equiv P\{N_\infty = n\}$, is more fruitful in the sense that it generally produces either explicit formulas or computationally tractable procedures to determine $\{p_n\}$, at least approximately. We will let $G(z) = \sum_n p_n z^n$ be the generating function of the steady-state N_∞.

The average number of customers in the system ($L \equiv E[N_\infty]$), is simply the mean value of the limiting distribution. One can also determine the average number of customers in the queue (L_q) by using this distribution. Thus, other performance measures, like the average waiting time in the system (W), and in the queue (W_q) can be computed as a direct consequence of Little's formulas, i.e., $W = L/\lambda_e$ and $W_q = L_q/\lambda_e$, where λ_e is the effective arrival rate. The server utilization factor u, or the expected number of busy servers under steady-state conditions, is another important measure that can generally be obtained through $\{p_n\}$.

2. ADVANCED MARKOV MODELS

2.1. General Approach

A queueing model is classified as Markovian if the process N is a Markov process on the state space $E = \{0, 1, 2, ...\}$. This makes it possible to utilize the existing theory on the transient and ergodic behavior of Markov processes. In particular, if A is the transition rate matrix, or generator, of N, then the so-called difference-differential equations of queueing theory are direct consequences of Chapman-Kolmogorov differential equations, or

$$\frac{d}{dt}P_n(t) = \sum_{m=0}^{\infty} P_m(t)A(m,n) , \qquad n \geq 0,\ t \geq 0 \tag{2}$$

One can use available methods, like moment generating functions or linear operators, to solve these equations. This represents the crust of transient analysis. In the ergodic case, the limiting distribution satisfies the so-called difference equations of queueing theory, which is the well-known result $pA = 0$, or

$$\sum_{m=0}^{\infty} p_m A(m,n) = 0 , \qquad n \geq 0 \tag{3}$$

It is relatively easier to solve these equations when compared with the transient case. The following queueing models illustrate how these results can be manipulated.

2.2. Markovian Queues With Bulk Arrivals ($M^B/M/1$)

In many applications, customers arrive to a queue in groups rather than singly. This will be the case when a boxful of parts arrive to a workstation for processing, or when an airplane, bus, or train arrives with many passengers on board. Here, A is a compound Poisson process with some arrival rate λ, and the group size, B, has some distribution $\{q_n\}$. If the service rate is μ, then the transition rate graph of A is as given in Figure 1.

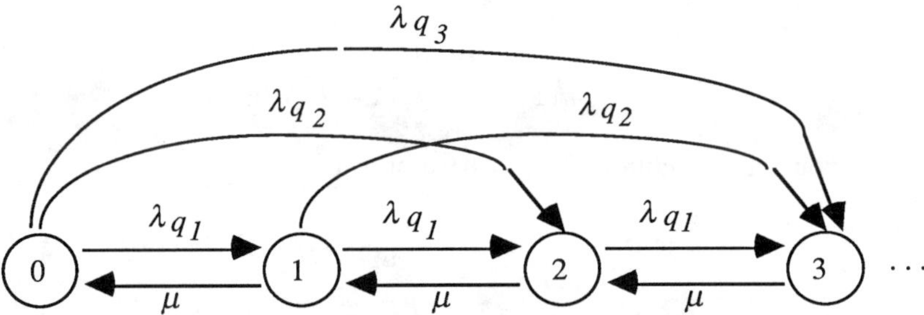

Figure 1. Transition Rate Graph of an $M^B/M/1$ Queue

Moreover, the difference-differential equations (2) become

$$\frac{d}{dt}P_n(t) = -(\lambda + \mu)P_n(t) + \mu P_{n+1}(t) + \lambda \sum_{k=1}^{n} P_{n-k}(t)q_k , \qquad n \geq 1 \tag{4}$$

$$\frac{d}{dt}P_0(t) = -\lambda P_0(t) + \mu P_1(t)$$

Under steady-state conditions, we obtain the well-known difference equations (3) as

$$(\lambda + \mu)p_n = \mu p_{n+1} + \lambda \sum_{k=1}^{n} p_{n-k} q_k \ , \qquad n \geq 1 \tag{5}$$

$$\lambda p_0 = \mu p_1$$

It is possible to solve for $\{p_n\}$ in (5) using generating functions in the following manner:

$$(\lambda + \mu) \sum_{n=0}^{\infty} p_n z^n - \mu p_0 = \frac{\mu}{z} \sum_{n=1}^{\infty} p_n z^n + \lambda \sum_{n=1}^{\infty} \sum_{k=1}^{n} p_{n-k} q_k z^n$$

$$= \frac{\mu}{z} \sum_{n=1}^{\infty} p_n z^n + \lambda \sum_{k=1}^{\infty} q_k z^k \sum_{n=k}^{\infty} p_{n-k} z^{n-k}$$

or equivalently,

$$(\lambda + \mu)G(z) - \mu p_0 = \frac{\mu}{z}[G(z) - p_0] + \lambda G(z)H(z) \tag{6}$$

where $H(z) = \sum q_k z^k$ is the generating function of B, and G is the generating function of N_∞. Equation (6) implies

$$G(z) = \frac{\mu p_0 (1 - z)}{\mu(1 - z) - \lambda z(1 - H(z))} \tag{7}$$

and using the boundary condition $G(1) = 1$ we obtain

$$G(1) = 1 = \frac{-\mu p_0}{-\mu + \lambda b} \tag{8}$$

where $b = E[B]$, i.e., the expected group size. Therefore, defining the traffic intensity as $\rho = \lambda b/\mu$, we have $p_0 = 1 - \rho$ and

$$G(z) = \frac{\mu(1 - \rho)(1 - z)}{\mu(1 - z) - \lambda z(1 - H(z))} \tag{9}$$

provided that $p_0 = 1 - \rho > 0$, or $\rho < 1$. One can easily determine the steady-state distribution using the fact that

$$p_n = \frac{1}{n!}\left[\frac{dG^n(z)}{dz^n}\right]\Big|_{z=0} \tag{10}$$

The procedure outlined above is a rather streamlined approach in analyzing any Markov queueing model; i.e., determine the difference or difference-differential equations through Chapman-Kolmogorov equations, and solve them using generating functions, Laplace-Stieltjes transforms, recursive relationships, linear operators, etc..

The special case with constant group size $B = b$ has $H(z) = z^b$, so that (9) reduces to

$$G(z) = \frac{\mu(1-\rho)}{\mu - \lambda z \sum_{j=0}^{b-1} z^j} \tag{11}$$

The expected number of customers in the system under steady-state conditions, L, can be obtained as

$$L = \left(\frac{dG(z)}{dz}\right)\Big|_{z=1} = \frac{\rho(1+b)}{2(1-\rho)} \tag{12}$$

Moreover, other performance measures are

$$\begin{aligned}
L_q &= \sum_{n=1}^{\infty}(n-1)p_n = L - (1 - p_0) = L - \rho \\
\lambda_e &= b\lambda \\
W &= \frac{L}{\lambda_e} = \frac{(1+b)}{2\mu(1-\rho)} \\
W_q &= \frac{L_q}{\lambda_e} = W - \left(\frac{1}{\mu}\right) \\
u &= 1 - p_0 = \rho\,.
\end{aligned} \tag{13}$$

2.3. Markovian Queues With Batch Service $(M/M^B/1)$

In some queueing applications, customers may be served in batches rather then one by one. This is usually the case in batch servicing computer systems, special workstations which process many items at the same time, or transportation systems which operate on a full-shuttle basis. The arrival process A is an ordinary Poisson process with rate λ, and each batch service is exponentially distributed with rate μ independent of the batch size. We assume that the server can handle a batch of at most B customers at a time for some fixed constant B. The transition rate graph is given in Figure 2.

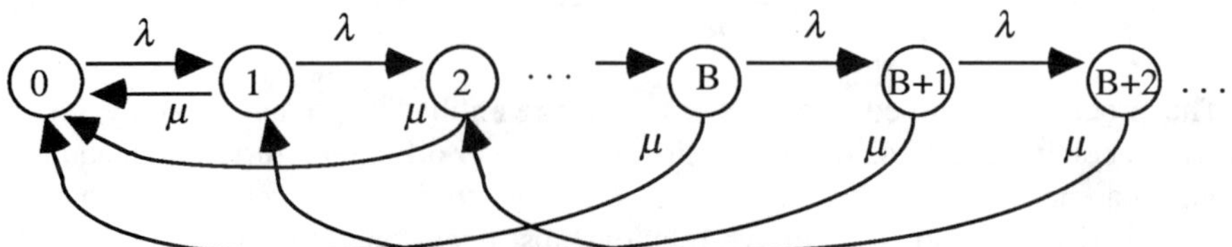

Figure 2. Transition Rate Graph of an $M/M^B/1$ Queue

The steady-state difference equations (3) corresponding to Figure 2 are easily seen to be

$$(\lambda + \mu)p_{n+1} = \lambda p_n + \mu p_{n+B+1} , \qquad n \geq 0 \tag{14}$$
$$\lambda p_0 = \mu(p_1 + p_2 + \cdots + p_B)$$

which can be written as

$$[\mu T^{B+1} - (\lambda + \mu)T + \lambda]p_n = 0 , \qquad n \geq 0 \tag{15}$$

where the operator T is defined as $T(p_n) = p_{n+1}$. Using this linear operator one can verify that the limiting distribution is of the form

$$p_n = (1 - r)r^n , \qquad n \geq 0 \tag{16}$$

where r is the smallest root in $[0, 1]$ of

$$\mu x^{B+1} - (\lambda + \mu)x + \lambda = 0$$

This implies that $L = r/(1-r)$, $W = r/\lambda(1-r)$, $L_q = r^B L$, $W_q = r^B W$, and $u = r$.

2.4. Markovian Queues With Priorities

The first-come first-served (FCFS) service discipline is a fair procedure in determining the order in which customers are served. However, this is not the case in many service systems and customers are classified according to different priorities. VIP, first-class, and economy-class priorities are almost always given to airline passengers. Users of computer systems are routinely given different priority levels to access the system and run their programs.

Suppose that all customer arrivals can be classified according to different priority classes indexed in $\{1, 2, ..., k\}$, so that $i < j$ implies that type-i customers have higher priority over type-j customers. Class i is said to have nonpreemptive (or head-of-the-line) priority over class j if an arriving type-i customer joins the queue at the head-of-the-line among type-j customers. Thus, the servicing of a type-j customer is not interrupted by the arrival of a type-i customer.

Class i is said to have preemptive priority over class j if an arriving type-i customer not only joins the queue at the head-of-the-line among type-j customers, but also displaces a type-j customer who may be in service. The servicing of a customer who is preempted from service may continue from the point of the interruption, this rule is called preemptive-resume. Or, the server may have to restart the service from the beginning; in which case we have the preemptive-repeat rule. If the duration of the repeated service is identical to the initial one, then this is called the preemptive-repeat-identical rule. Otherwise, if the repeated service duration is different, then the priority rule is called preemptive-repeat-different.

As an illustration, consider a model with preemptive-resume-different priority rule and only 2 types of customer arrivals. Assume that the two arrival processes, A^1 and A^2, are independent Poisson processes with respective rates λ_1 and λ_2. Furthermore, the service duration of type-1 and type-2 customers are exponentially distributed with rates μ_1 and μ_2, respectively. Taking $N = (N^1, N^2)$ as a vector process where N_t^1 and N_t^2 represent the number of type-1 and type-2 customers, respectively, in the system at time t, we observe that N is a Markov process with a transition rate graph given in Figure 3.

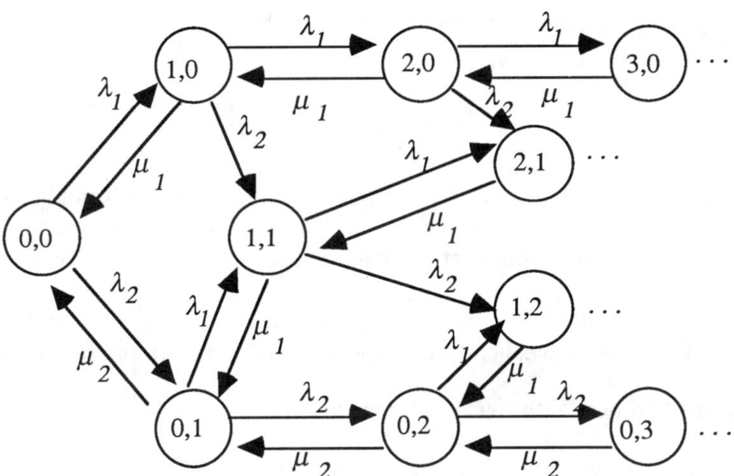

Figure 3. Transition Rate Graph of an $M/M/1$ Queue with 2 Priority Classes

It is clear that the system formed by type-1 customers is an independent $M/M/1$ queue so that $p_n^1 = (1 - \rho_1)\rho_1^n$, $L^1 = \rho_1/(1 - \rho_1)$, and $W^1 = 1/\mu_1(1 - \rho_1)$ where $\rho_1 = \lambda_1/\mu_1$. The steady-state equations (3) using the transition rate matrix A given by

Figure 3 can be written as

$$p_{m,n}(\lambda_1 + \lambda_2 + \mu_1 1(m) + \mu_2 1(n)(1 - 1(m))) \qquad (17)$$
$$= \lambda_1 p_{m-1,n} + \lambda_2 p_{m,n-1} + \mu_1 p_{m+1,n} + \mu_2(1 - 1(m))p_{m,n+1} , \qquad n, m \geq 0$$

where $1(x)$ is the indicator function which is equal to one if $x \neq 0$, and zero otherwise. The generating function $G(z,y) = \sum_{m,n} p_{m,n} z^m y^n$ can be determined to be

$$G(z,y) = \left(\frac{1 - \rho_1 - \rho_2}{1 - \eta - y\rho_2}\right)\left(\frac{1 - \eta}{1 - \eta z}\right) \qquad (18)$$

where $\rho_2 = \lambda_2/\mu_2$ and

$$\eta = \frac{1}{2\mu_1}(\mu_1 + \lambda_1 + \lambda_2(1 - y) - \sqrt{(\mu_1 + \lambda_1 + \lambda_2(1 - y))^2 - 4\lambda_1\mu_1}) . \qquad (19)$$

The limiting distribution $\{p_{n,m}\}$ can be found from (18); and in particular, $p_{0,0} = 1 - \rho_1 - \rho_2$, $u = \rho_1 + \rho_2$

$$L_1 = \rho_1/(1 - \rho_1) \qquad (20)$$
$$L_2 = \left(\frac{\rho_2}{1 - \rho_1 - \rho_2}\right)\left(1 + \left(\frac{\mu_2}{\mu_1}\right)L^1\right)$$
$$L_q^1 = \rho_1^2/(1 - \rho_1)$$
$$L_q^2 = \left(\frac{\rho_2}{1 - \rho_1 - \rho_2}\right)\left(\rho_1 + \rho_2 + \left(\frac{\mu_2}{\mu_1}\right)L^1\right)$$

and the average waiting times can be computed through Little's formula in a trivial manner. Finally, note that the results presented by (17)-(20) apply also for preemptive-resume-identical, or preemptive-repeat cases because of the memorylessness property of the exponential distribution.

Queues with priorities were introduced in Cobham (1954), and the textbook by Jaiswal (1968) presents an excellent account of most of the work on the subject.

2.5. Markovian Queues With Service Interruptions

In most queueing systems, it is not possible to keep the server operational at all times, and service can thus be interrupted. The main reason for this is the failure or breakdown of the server. There can also be scheduled service interruptions during weekends or vacations, at nights, etc.. In priority queues, the arrival of a higher priority customer may cause an interruption in the service of a lower priority customer if the service discipline is preemptive. As a matter of fact, queues with service interruptions can be modeled and analyzed as priority queues with two types of customer arrivals. Here, the arrival of a higher priority customer represents a breakdown or service interruption, and causes the start of a so-called off-period. Thus, service duration of such customers are the durations of off-periods, and on-periods start as soon as off-periods end (i.e., higher priority customer is served) provided that there are no higher priority customers in the queue.

Consider, as an example, the following single-server Markovian model where customers arrive according to a Poisson process with rate λ, and services are exponential with rate μ. Moreover, assume that the server fails after an exponential amount of time with rate f, and each repair or maintenance sojourn is also exponentially distributed with rate r. Letting $Z = \{Z_t; t \geq 0\}$ be an indicator process which is equal to 0 or 1 accordingly if the server is operational or in repair respectively, one observes that the process (N, Z) is a Markov process with the transition rate graph given in Figure 4.

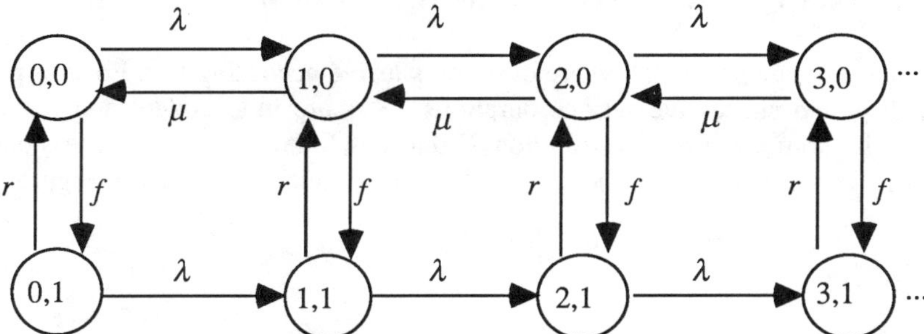

Figure 4. Transition Rate Graph of an $M/M/1$ Queue with Service Interruptions.

This model was first introduced and analyzed in White and Christie (1958) using similar procedures as in the previous sections, and the generating function $G(z, y)$ for the steady-state distribution of (N, Z) is reported to be

$$G(z, y) = (1 - u)\left\{1 + \frac{1 - \mathcal{L}_1(\lambda(1 - z))}{\lambda(1 - z)} \, fy\right\} \tag{21}$$

$$\left\{1 - \frac{1 - \mathcal{L}_2(f(1 - \mathcal{L}_1(\lambda(1 - z))) + \lambda(1 - z))}{1 - (1/z)\mathcal{L}_2(f(1 - \mathcal{L}_1(\lambda(1 - z))) + \lambda(1 - z))}\right\}$$

where $\mathcal{L}_1(x) = (r/(r + x))$, $\mathcal{L}_2(x) = (\mu/(\mu + x))$ are the Laplace-Stieltjes transforms of repair durations and service durations respectively, and u is the server utilization factor with $u = (\lambda/\mu) + (f/(r + f))$. Moreover, one can easily obtain from (21) that

$$\mathrm{L} = \lambda\big((r + f)^2 + \mu f\big)\Big/\big((r + f)(r(\mu - \lambda) + \lambda f)\big) \tag{22}$$

This simple model in White and Christie (1958) initiated a whole literature on queueing models with service interruptions since its appearance. The interested reader is referred to Federgruen and Green (1986) for a rather complete account of the literature. The generating function (21) is generalized in Thiruvengadam (1963) to the case where both repair and service durations are arbitrary with Laplace-Stieltjes transforms $\mathcal{L}_1$ and $\mathcal{L}_2$, respectively. An all-exponential model where the arrival and service rates vary during on and off-periods is considered in Eisen and Tainiter (1963). Avi-Ithzak and Naor (1963) study five models with different assumptions on the service and failure structures. The $M/M/N$ queueing system with server breakdowns is analyzed in Mitrany and Avi-Ithzak (1968), as well as in Neuts and Lucantoni (1979).

3. SEMI-MARKOV QUEUEING MODELS

3.1. Semi-Markov Queues With General Service Distribution ($M/G/1$)

Consider a queueing model where customers arrive according to a Poisson process A with rate λ, and successive service durations $\{S_n\}$ are independent and identically distributed with some arbitrary distribution F that has finite mean $1/\mu$. The stochastic evolution of the number of customers present in the system is pictured in Figure 5.

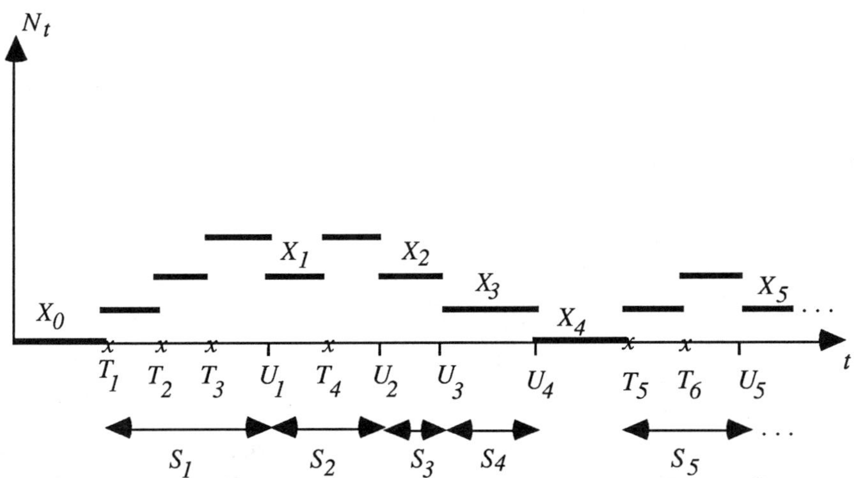

Figure 5. The Stochastic Evolution of N in an $M/G/1$ Queue

It is clear that N is no longer a Markov process, and the streamlined approach employed in the previous section can no longer be utilized. Although it is difficult to deal with N directly, one can use the so-called "embedded Markov chain" to analyze N. This method, introduced in Kendall (1953), is based on the observation that the number of customers left in system at departure epochs form a Markov chain.

It is clear from Figure 5 that the process $\{X_n \equiv N(U_n)\}$ is embedded in N, and that it is a Markov chain with transition matrix

$$
P = \begin{pmatrix}
q_0 & q_1 & q_2 & q_3 & \cdot & \cdot & \cdot & \cdot \\
q_0 & q_1 & q_2 & q_3 & \cdot & \cdot & \cdot & \cdot \\
0 & q_0 & q_1 & q_2 & \cdot & \cdot & \cdot & \cdot \\
0 & 0 & q_0 & q_1 & \cdot & \cdot & \cdot & \cdot \\
\cdot & \cdot & \cdot & \cdot & \cdot & \cdot & \cdot & \cdot \\
\cdot & \cdot & \cdot & \cdot & \cdot & \cdot & \cdot & \cdot \\
\cdot & \cdot & \cdot & \cdot & \cdot & \cdot & \cdot & \cdot
\end{pmatrix}
\tag{23}
$$

where q_k is the probability that there are n arrivals during a typical service, i.e.,

$$
q_k = P\{A_S = k\} = \int_0^\infty \frac{e^{-\lambda t}(\lambda t)^k}{k!}\, dF(t)\,, \qquad k = 0,1,\dots\,.
\tag{24}
$$

The process $\{(X_n, U_n)\}$ is a Markov renewal process, and N is a semi-regenerative process on it. We refer the reader to Çınlar (1975) for an excellent coverage of Markov renewal theory. One can use Markov renewal theory, as in Gelenbe and Pujolle (1987), to arrive at the interesting conclusion that the steady-state distribution of N is actually identical to the steady-state distribution of the Markov chain X. This implies that the distribution $\{p_n\}$ satisfies

$$
p_n = \sum_{m=0}^\infty p_m P(m,n)\,,
\tag{25}
$$

with $\sum_n p_n = 1$, or

$$
\begin{aligned}
p_0 &= p_0 q_0 + p_1 q_0 \\
p_1 &= p_0 q_1 + p_1 q_1 + p_2 q_0 \\
p_2 &= p_0 q_2 + p_1 q_2 + p_2 q_1 + p_3 q_0
\end{aligned}
\tag{26}
$$

$$
\cdot
$$
$$
\cdot
$$
$$
\cdot
$$

$$
1 = p_0 + p_1 + p_2 + \cdots
$$

The Markov chain X is ergodic if $\rho = \lambda/\mu < 1$, and (26) can be used in a trivial manner to obtain the generating function $G(z)$ of N_∞ as

$$
G(z) = \frac{(1-\rho)(z-1)\mathcal{F}(\lambda - \lambda z)}{z - \mathcal{F}(\lambda - \lambda z)}
\tag{27}
$$

where $\mathcal{F}$ is the Laplace-Stieltjes transform of F. Also, the server utilization factor is $u = 1 - p_0 = \rho$, and the average number of customers in the system is

$$L = \rho + \rho^2 \frac{1 + \mu^2 \sigma_\mu^2}{2(1 - \rho)} \qquad (28)$$

where σ_μ^2 is the variance of service durations. Equations (27) and (28) are known as Pollaczek-Khintchine formulas. It is clear that for given arrival and service rates, L is minimized if $\sigma_\mu^2 = 0$, or if service durations are deterministic. Moreover, $L_q = L - \rho$, and W and W_q can be computed through Little's formula to obtain other versions of the Pollaczek-Khintchine formulas.

The basic idea is analyzing M/G type queues is to employ the "embedded Markov chain" method as illustrated above. Even in more complicated applications, the basic process of interest, N, will no longer be a Markov process, but a semi-Markov process where the Markov property holds at customer departure times. Thus, analysis of the Markov chain embedded in N at these departure times provide useful results which can be reflected on N.

If the customer arrival process A is compound Poisson so that bulk arrivals are possible, then this $M^B/G/1$ queue can be analyzed in a similar fashion as the $M/G/1$ queue. Note that the queue process at departure epochs still form an embedded Markov chain, but now the transition matrix must be adjusted properly so that

$$q_k = P\{A_s = k\} = \int_0^\infty P\{A_t = k\}dF(t) \qquad k = 0, 1, \dots \, . \qquad (29)$$

The use of the "embedded Markov chain" method for an $M/G/1$ queue with service interruptions is demonstrated in Federgruen and Green (1988). Here, the durations of off-periods are independent and identically distributed with an arbitrary distribution, but on-period durations have a phase-type distribution. The queue process at departure times is no longer a Markov chain since the residual on-period distribution is arbitrary.

Phase-type distributions are characterized as absorption time distributions of finite state Markov processes. Thus, the residual on-period distribution is the same at all distinct time epochs, given that the phase process is in the same state. Calling this phase process Y, it follows now that $\{N(U_n), Y(U_n)\}$ is a Markov chain embedded in $\{N_t, Y_t\}$. This allows one to utilize Markov renewal theory as in ordinary $M/G/1$ queues, and characterize the steady-state distribution of N through that of the embedded Markov chain. Approximate and exact computational procedures are provided in Federgruen and Green (1986,1988), as well as light and heavy traffic approximations. The importance of phase-type distributions is also due to the fact that the class of phase-type distributions is dense in the class of continuous distributions as demonstrated in Neuts (1981).

Another line of research, closely related with service interruption models, assumes that the server takes a vacation or rest of random duration T_0 with some distribution function V whenever he becomes idle, or finds an empty queue upon his return from a vacation. This model was first analyzed in Miller (1964), and a literature review can be found in Doshi (1986) and Tejhen (1986). Vacation models are especially fit for queueing applications in telecommunication or data processing systems where a "vacation" implies that the server is unavailable because he is either searching for new work, serving another class of jobs, or doing maintenance.

The usual approach in analyzing vacation models involves, once again, the "embedded Markov chain" method with a slightly extended state space. This time we keep track of not only the number of customers in the system at departure and vacation completion times, but also an indicator process which shows if the embedding times are departure or vacation completion times. An interesting set of decomposition results are given in Scholl and Kleinrock (1983). For example, the generating function of the steady-state distribution has the following somewhat extended form of the Pollaczek-Khintchine formula

$$G(z) = \frac{(1-\rho)\mathcal{F}(\lambda - \lambda z)}{z - \mathcal{F}(\lambda - \lambda z)} \frac{1 - \mathcal{V}(\lambda - \lambda z)}{E[T_0](\lambda - \lambda z)} \tag{30}$$

where $\mathcal{V}$ is the Laplace-Stieltjes transform of V. It is clear that the first term on the right hand side of (30) is identical to the one given in (27), and the second term can be shown to be the Laplace-Stieljes transform of the number of customer arrivals during a time interval distributed as the stationary forward recurrence time of a rest period.

The characterization implied by (30) is that the number in the system for the vacation model behaves as the sum of two independent random variables: number in the system for a regular $M/G/1$ queue without vacation, and number of arrivals during a time interval distributed as the stationary forward recurrence time of the rest period. This implies, in particular, that the Pollaczek-Khintchine formula (28) now extends as

$$L = \rho + \frac{\rho^2(1 + \mu^2 \sigma_\mu^2)}{2(1-\rho)} + \lambda \frac{E[T_0^2]}{2E[T_0]}. \tag{31}$$

This decomposition is not only true for the number of customers in the system, but for the time spent in the system as well. This is demonstrated in Scholl and Kleinrock (1983) by a result which states that, under the FCFS discipline, the time spent in the system for the vacation model behaves as the sum of two independent random variables: time spent in the system for the regular $M/G/1$ queue without vacations, and an additional delay distributed as the stationary forward recurrence time of the rest period. These decomposition results are later generalized in Fuhrmann and Cooper (1985) for other models with different vacation structures.It is reported recently in Shantikumar (1988) that the stochastic decomposition property holds for the number of customers in the system in models where bulk arrivals are possible, as well as some variations involving reneging, balking, and arrival rates that are dependent on system state.

3.2. Semi-Markov Queues With General Arrivals $(G/M/1)$

This is the model where customer interarrival times are independent and identically distributed with some arbitrary distribution that has finite mean $1/\lambda$, and service durations are independent and exponentially distributed with rate μ. Figure 6 presents a typical sample-path of the queue process N.

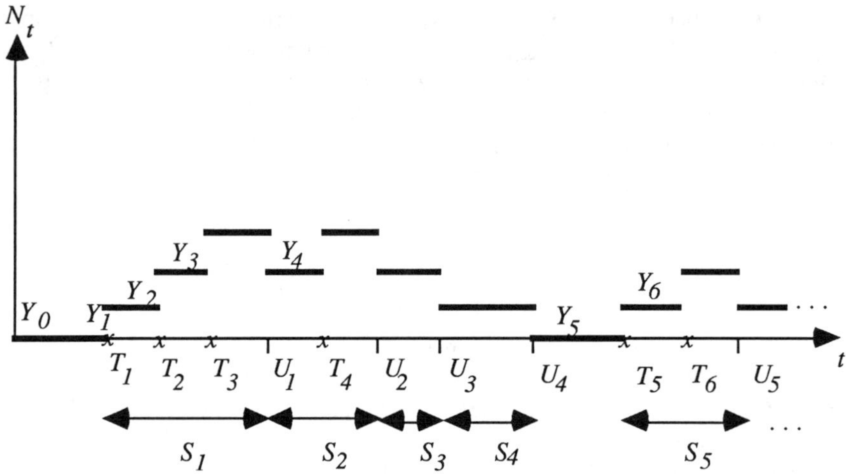

Figure 6. The Stochastic Evolution of N in a $G/M/1$ Queue

The process N is no longer a Markov process since the Markov property is violated due to the non-exponentiallity of the interarrival distribution. However, we can once again identify an embedded Markov chain $\{Y_n\}$ as shown in Figure 6. Now, $Y_n = N(T_{n-})$ is the number of customers in the system just before the nth arrival and its transition matrix is given by

$$P = \begin{pmatrix} r_0 & q_0 & & & & \\ r_1 & q_1 & q_0 & & & \\ r_2 & q_2 & q_1 & q_0 & & \\ \cdot & \cdot & \cdot & \cdot & \cdot & \\ \cdot & \cdot & \cdot & \cdot & \cdot & \cdot \\ & \cdot & \cdot & \cdot & \cdot & \cdot & \cdot \end{pmatrix} \tag{32}$$

where q_k is the probability that the number of services completed in between two arrivals is equal to k defined similarly as in (24), and $r_k = q_{k+1} + q_{k+2} + \cdots$.

The Markov chain Y is ergodic if $\rho = \lambda/\mu < 1$, and its steady-state distribution π can be computed using the system of linear equations $\pi = \pi P$, $\sum_n \pi_n = 1$ as

$$\pi_k = (1-\beta)\beta^k \qquad k = 0,1,2,\dots \tag{33}$$

where β is the unique solution in $(0,1)$ of

$$\beta = \sum_{k=0}^{\infty} q_k \beta^k \tag{34}$$

Contrary to the $M/G/1$ case, the steady-state distributions of the Markov chain Y and semi-Markov process N are not identical. However, the process $\{(Y_n, T_n)\}$ is a Markov renewal process, and N is a semi-regenerative process on it. Using Markov renewal theory as in Gelenbe and Pujolle (1987), the steady-state distribution $\{p_n\}$ can be found to be

$$p_0 = 1 - \rho \tag{35}$$
$$p_k = \rho(1-\beta)\beta^{k-1} \qquad k = 1,2,\dots$$

provided that $\rho < 1$. This implies that $L = \rho/(1-\beta)$, and $W = 1/\mu(1-\beta)$ through Little's formula.

Letting $V_n(t)$ denote the distribution function of the waiting time in the queue of the nth arriving customer; then, given Y_n, $V_n(t)$ is the Y_n-fold convolution of the exponential distribution with rate μ. This implies that

$$\lim_{n \to \infty} V_n(t) = \sum_{n=0}^{\infty} (1-\beta)\beta^n P\{S_1 + S_2 + \cdots + S_n \le t\} \tag{36}$$
$$= (1-\beta) + \sum_{n=1}^{\infty} (1-\beta)\beta^n \int_0^t \frac{\mu(\mu s)^{n-1} e^{-\mu s}}{(n-1)!}\, ds$$
$$= 1 - \beta e^{-\mu(1-\beta)t}, \qquad t \ge 0$$

so that the mean is $W_q = \beta/\mu(1-\beta)$, and $L_q = \rho\beta/(1-\beta)$.

Although the literature on M/G type queues is quite abundant as illustrated by various applications in the previous section, G/M type queues have not attracted as much interest from researchers. This is undoubtedly due to the fact that the Poisson process is very adequate to model the customer arrival process in many applications.

4. OTHER QUEUEING MODELS

4.1. Non-Markov Queues With General Service and Arrival Distributions $(G/G/1)$

Consider a single-server queueing model where customer interarrivals $\{T_{n+1} - T_n\}$ are independent and identically distributed with an arbitrary distribution that has mean $1/\lambda$ and variance σ_λ^2, and customer services $\{S_n\}$ are also independent and identically distributed with an arbitrary distribution that has mean $1/\mu$ and variance σ_μ^2. A typical sample path of the queue process N is presented in Figure 6. This time, however, we can not identify an embedded Markov chain because of the nonexponentiality of the distributions of both services and interarrivals.

The usual approach in analyzing these models concentrates on the delay process $\{D_n\}$ where D_n is the delay in the servicing of the nth customer, or the amount of the time the nth arriving customer waits in the queue. It is well-known that

$$D_{n+1} = (D_n + S_n - (T_{n+1} - T_n))^+ , \qquad n = 0, 1, ... \qquad (37)$$

where $a^+ = max\{a, 0\}$ for any real number a. Therefore, $\{D_n\}$ is a Markov chain. It is demonstrated in Cobham (1954) that $\{D_n\}$ is ergodic if $\rho = \lambda/\mu < 1$. Moreover, it follows from (37) that the distribution function $D_n(t) = P\{D_n \leq t\}$ satisfies

$$D_{n+1}(t) = \int_{-\infty}^{t} D_n(t - x)dH(x) , \qquad t \geq 0 \qquad (38)$$

where $H(t) = P\{S_n - (T_{n+1} - T_n) \leq t\}$. If $\rho < 1$, then $D = \lim_{n \to \infty} D_n$ satisfies the so-called Lindley equation

$$D(t) = \int_{-\infty}^{t} D(t - x)dH(x) , \qquad t \geq 0 \qquad (39)$$

which can not be solved expilicitly except for some special cases. In the $G/M/1$ case, for example, the solution to this equation is given by (36).

For arbitrary G/G type models, it is generally not possible to determine performance measures like L, L_q, W, and W_q exactly. However, one can obtain approximations and bounds. The equality (37) is exploited in Gelenbe and Pujolle (1987), to arrive at the approximation.

$$L \cong \rho + \rho^2 \frac{\lambda^2\sigma_\lambda^2 + \mu^2\sigma_\mu^2}{2(1 - \rho)} \qquad (40)$$

which coincides with the Pollaczek-Khintchine formula (28) in the $M/G/1$ case where $\sigma_\lambda^2 = 1/\lambda^2$. This approximation was first obtained in Kingman (1965); and the reader is referred to Stoyan (1977) for a survey of various bounds and approximations.

The difficulty in studying the Lindley equation (39) leads researchers to use other modeling tools like virtual-delay processes and diffusion processes.

4.2. Virtual-Delay Processes

We have thus far concentrated on the queue process N in order to evaluate various measures of effectiveness. Figures 5 and 6 are examples along this direction. It is possible to represent the sample-path information in these figures by defining another process, called the virtual-delay process $\{W_t\}$, so that Wt is the amount of work the server is facing at time t. This definition justifies the term "virtual-delay" since W_t is also the amount of time a customer who arrives at time t will wait under the FIFO discipline. A sample-path of $\{W_t\}$ corresponding to Figure 6 is presented in Figure 7.

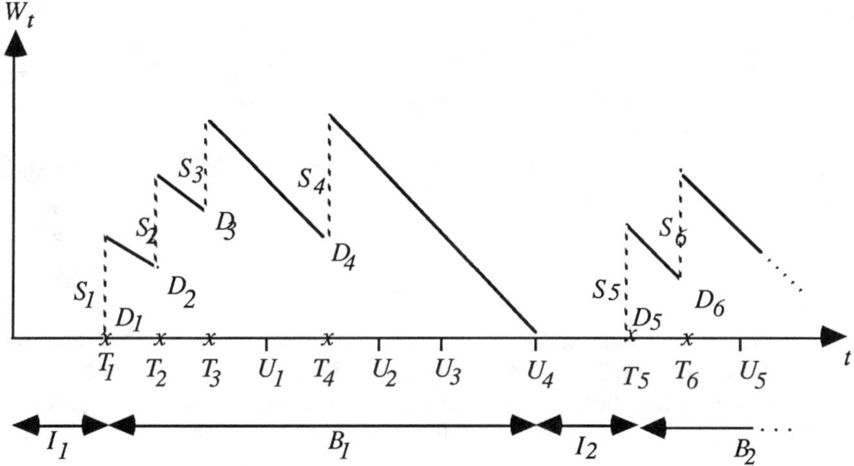

Figure 7. The Sample-Path of the Virtual-Delay Process Corresponding to Figure 6.

Here, $\{I_n\}$ and $\{B_n\}$ represent the duration of consecutive idle and busy cycles, respectively. The waiting time of the nth arriving customer is $D_n = W(T_{n-})$ as depicted in Figure 7. It is clear that, under independent interarrival and service durations assumption, $\{D_n\}$ is a Markov chain and it satisfies (37). Therefore, the process $\{W_t\}$ is a semi-Markov process where the Markov property holds at its stopping times $\{T_n\}$.

The characterizations provided in the previous section imply that $D_n \to D$ (i.e., D_n converges in distribution to a random variable D), and $D \sim (D + S - A)^+$ where S and A are two generic random variables representing service and interarrival durations, respectively.

The description of the $G/G/1$ queue using virtual-delay processes is fully exploited in Doshi (1985) in analyzing the server-vacation model where the server is allowed to take vacations of random durations whenever he becomes idle, or finds an empty queue upon his return. The analysis is based on a sample-path comparison of the virtual-delay processes of a regular $G/G/1$ queue and a $G/G/1$ queue with vacations as shown in Figure 8.

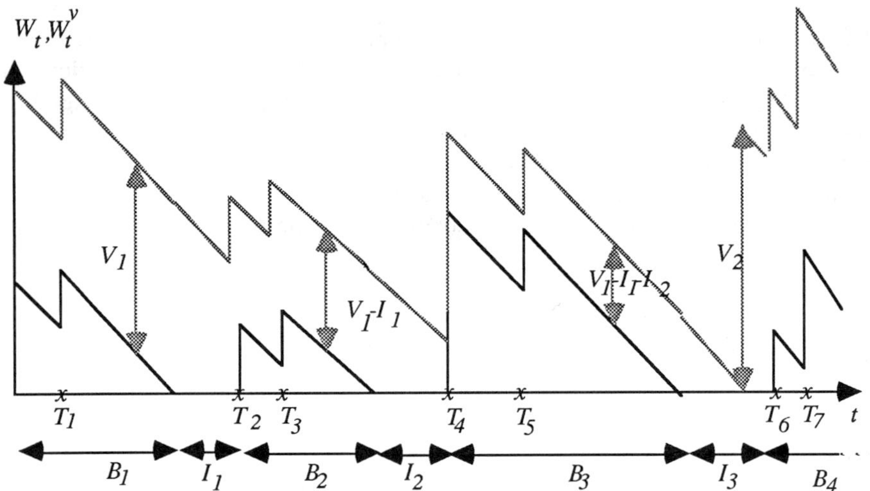

Figure 8. Sample-Path of Virtual-Delay Processes for the Regular $G/G/1$ and Server-Vacation Models

Here, $\{W_t\}$ is the virtual-delay process of the regular $G/G/1$ queue represented by the solid lines, and $\{W_t^v\}$ is the virtual-delay process of server vacation model represented by the dashed lines. The vacation durations are given by $\{V_n\}$, and both sample-paths are drawn using a given realization of the customer arrival and service processes.

Denoting by $\{D_n\}$ and $\{D_n^v\}$ the corresponding waiting time of the nth customer for the regular $G/G/1$ and server-vacation models respectively, it is clear from Figure 8 that at all arrival times that fall in some busy cycle B_k, we have

$$D_n^v = D_n + Y_k , \qquad T_n \in B_k \tag{41}$$

where Y_k is the constant difference between W_t^v and W_t over B_k. This leads to the stochastic decomposition property for the server-vacation model which states that if $D_n \to D$, then $D^v = \lim_{n \to \infty} D_n^v$ satisfies

$$D^v = D + Y \tag{42}$$

where D and Y are independent random variables. Here, D and D^v are tne waiting times of an arbitrary customer under steady-state conditions for the regular $G/G/1$ and server-vacation models respectively. Moreover, Y is a generic random variable representing the stationary forward recurrence time of the vacations; i.e.,

$$P\{Y > y\} = \frac{\int_y^\infty P\{V_1 > x\}dx}{E[V_1]} \; , \qquad y \geq 0 \qquad (43)$$

A similar stochastic decomposition result is provided in Doshi (1985) for the $G/G/1$ queue with set-up time, where the server spends a random set-up time before starting the first service in each busy cycle.

4.3. Diffusion Approximations

Another modeling tool often used in analyzing complex queueing models is the approximation of the process of interest, like the number of customers in the system or the virtual-delay process, by a diffusion process. This tool is generally employed for G/G type queues because the other types can be analyzed using embedded Markov chains as demonstrated in the previous section. Diffusion approximations of queueing processes were first used in Gaver (1968), and Iglehart (1965).

Consider a $G/G/1$ queue with general independent services with mean $1/\mu$ and variance σ_μ^2, and general independent interarrivals with mean $1/\lambda$ and variance σ_λ^2. It is clear that the queue process N satisfies

$$N_t = N_0 + A_t - D_t \; , \qquad t \geq 0 \qquad (44)$$

where A and D are the customer arrival and departure processes respectively, and N_0 is the initial number of customers in the system.

Note that customer arrival epoches form a renewal process and it is well-known in renewal theory that the renewal counting process A satisfies

$$E[A_t] \cong \lambda t \; , \quad \mathrm{Var}(A_t) \cong \lambda^3 \sigma_\lambda^2 t \qquad (45)$$

The departure process D depends on A, and is not a renewal counting process in general. If $\rho = \lambda/\mu$ is close to one, however, we can approximate D by a renewal counting process so that

$$E[D_t] \cong \mu t \; , \quad \mathrm{Var}(D_t) \cong \mu^3 \sigma_\mu^2 t \qquad (46)$$

This leads to the approximation of N by a diffusion process $\mathcal{N}$ which has drift rate a and diffusion coefficient b^2 given by

$$a = \lambda - \mu \; , \quad b^2 = \lambda^3 \sigma_\lambda^2 + \mu^3 \sigma_\mu^2 \; . \qquad (47)$$

The fact that the number of customers in the system can not be negative requires that an adequate boundary condition should be imposed on $\mathcal{N}$. So, we assume that $\mathcal{N}$ is reflected at the boundary state 0. A diffusion process has continuous sample-paths which are nowhere differentiable, and have unbounded variations. Although it is impossible to draw a realization of such a process, Figure 9 attempts to provide a rough picture.

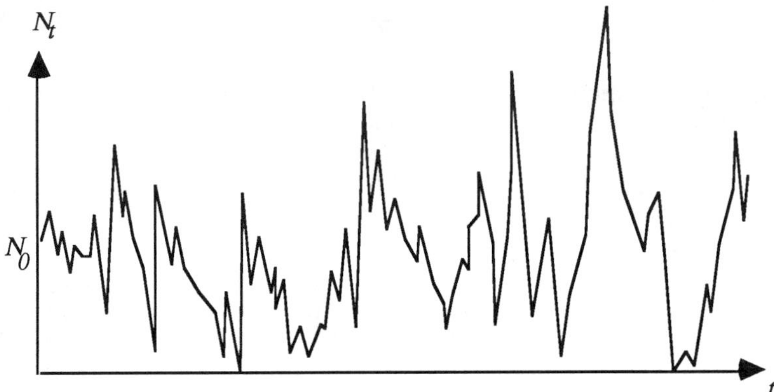

Figure 9. A Rough Representation of the Sample-Path of the Approximation $\mathcal{N}$

The probability law of $\mathcal{N}$ satisfies the so-called diffusion equation or Fokker-Planck equation

$$\frac{d}{dt}f(x,y,t) = -a\frac{d}{dt}f(x,y,t) + \frac{b^2}{2}\frac{d^2}{dy^2}f(x,y,t) \qquad (48)$$

where

$$f(x,y,t) = \frac{d}{dy}P\{\bar{N}_t \leq y|\bar{N}_0 = x\} \qquad (49)$$

If $\mathcal{N}$ is not reflected at state 0, then it is well-known that it can be represented as

$$\mathcal{N}_t = N_0 + at + bB_t , \qquad t \geq 0 \qquad (50)$$

where B is the standard Brownian motion, or the Wiener process. In other words, B is a time-homogeneous continuous Markov process on the real-line which satisfies

$$P\{B_t \leq z\} = \Phi(z) \qquad (51)$$

where Φ is the standard Normal distribution function; i.e.,

$$\Phi(z) = \frac{1}{\sqrt{2\pi}} \int_{-\infty}^{z} e^{-u^2/2} du . \qquad (52)$$

This leads to the probability law

$$P\{\mathcal{N}_t \leq y | \mathcal{N}_0 = x\} = \Phi\left(\frac{y - x - at}{b\sqrt{t}}\right) \qquad (53)$$

for the Brownian motion $\mathcal{N}$ if it is not forced to reflect at state 0.

On the other hand, if $\mathcal{N}$ is reflected at state 0, as it is in our approximation, then we have

$$P\{\mathcal{N}_t \leq y | \mathcal{N}_0 = x\} = \Phi\left(\frac{y - x - at}{b\sqrt{t}}\right) - e^{2xa/b^2}\Phi\left(\frac{-y - x - at}{b\sqrt{t}}\right) \qquad (54)$$

The reader can find a derivation of (54) in Newell (1974). The steady-state distribution of $\mathcal{N}$ exists provided that $a = \lambda - \mu < 0$, or $\rho = \lambda/\mu < 1$, so that

$$\lim_{t \to \infty} P\{\mathcal{N}_t \leq y | \mathcal{N}_t = x\} = 1 - e^{-(2a/b^2)y} , \qquad y \geq 0 \qquad (55)$$

independent of the initial state x. This implies that the expected number of customers in the system under steady-state conditions can be approximated by

$$L \cong \lim_{t \to \infty} E[\mathcal{N}_t | \mathcal{N}_0 = x] = \frac{-b^2}{2a} \qquad (56)$$

which can be rewritten as

$$L \cong \frac{\rho\lambda^2\sigma_\lambda^2 + \mu^2\sigma_\mu^2}{2(1 - \rho)} \qquad (57)$$

Since the steady-state N_∞ takes on discrete values, it is also possible to obtain various approximations for the steady-state distribution $\{p_n\}$ by discretizing the continuous distribution (55). Some possible procedures are outlined in Gelenbe and Pujolle (1987). In particular, one can take

$$p_n \cong \int_{n-1}^{n} (2\frac{a}{b^2})e^{-(2a/b^2)y} dy \qquad (58)$$

which gives

$$p_0 \cong 1 - \rho \tag{59}$$

$$p_n \cong [\rho(\rho\lambda^2\sigma_\lambda^2 + \mu^2\sigma_\mu^2)/2(1-\rho)](e^{n(2a/b^2)}(1 - e^{-2a/b^2})) , \qquad n \geq 1$$

so that we get the approximation

$$L \cong \rho + \frac{\rho(\rho\lambda^2\sigma_\lambda^2 + \mu^2\sigma_\mu^2)}{2(1-\rho)} \tag{60}$$

Note that approximation (60) is similar to the Pollaczek-Khintchine formula (28) in the $M/G/1$ case, since this implies $\lambda^2\sigma_\lambda^2 = 1$.

Another interesting approximation suggested in Gelenbe and Pujolle (1987) is to take

$$a = \lambda - \mu , \qquad b^2 = \lambda^2\sigma_\lambda^2 + \mu^2\sigma_\mu^2\rho \tag{61}$$

in the diffusion approximation, since there can be departures from the system only a proportion ρ of the time. This time, the discretization procedure explained by (58) gives the same distribution (59) with the new b^2 value. Thus, (60) now becomes

$$L \cong \rho + \frac{\rho^2(\lambda^2\sigma_\lambda^2 + \mu^2\sigma_\mu^2)}{2(1-\rho)} \tag{62}$$

which is the classical approximation of Kingman. Moreover, this is identical to the Pollaczek-Khintchine formula (28) in the $M/G/1$ case.

One can also approximate the virtual-delay process $\{W_t\}$ by a diffusion process $\{\mathcal{W}_t\}$ with drift rate c and diffusion coefficient d^2 given by

$$c = \rho - 1 , \qquad d^2 = \lambda(\sigma_\mu^2 + \rho^2\sigma_\lambda^2) . \tag{63}$$

The reader can find a demonstration of this approximation in Heyman and Sobel (1982). Since the virtual-delay can not be negative, $\{\mathcal{W}_t\}$ should be reflected at state 0. Therefore, the conditional and steady-state distributions of $\{\mathcal{W}_t\}$ are given by (54) and (58) where a and b^2 are replaced by c and d^2, respectively. Moreover, an approximation for W_q can be obtained using the steady-state distribution as

$$W_q \cong \frac{-d^2}{2c} = \lambda\frac{\sigma_\mu^2 + \rho^2\sigma_\lambda^2}{2(1-\rho)} \tag{64}$$

which is identical, once again, to the Pollaczek-Khintchine formula for W_q in the $M/G/1$ case.

In approximating a process by a diffusion process, one must first choose or determine the drift and diffusion functions $a(t)$ and $b^2(t)$ which can necessarily depend on time in general. If they do not, as in the examples mentioned above, the approximating process is actually a Brownian motion. Moreover, the boundary conditions that must be imposed on the diffusion equation also play an important role. In queueing applications, the process is either made reflecting at state 0, or it is made to jump instantaneously to some other state, like state 1. Finally, once the transient or steady-state distribution of the approximating process is determined, the practitioner is faced with the problem of discretizing this continuous distribution to obtain approximate measures associated with the real process being analyzed.

Diffusion process approximations can also be made for queueing models which are more general than the ones mentioned in this section. As examples, the reader can refer to Halachami and Franta (1978) for multiserver queueing systems, and Kobayashi (1974) for queueing networks. An approximation to queueing systems with interruptions can be found in Fischer (1977).

5. REFERENCES

1. Avi-Ithzak, B., and Naor, P. 1963. Some Queueing Problems with the Service Station Subject to Breakdown. *Opns. Res.* **11**, 303-320.

2. Bhat, U.N. 1969. Sixty-Years of Queueing Theory. *Mgmt. Sci* **15**, B280-B294.

3. Cobham, A. 1954. Priority Assignment in Waiting Line Problems. *Opns. Res.* **2**, 70-77.

4. Cohen, J.W. 1969. *The Single Server Queue.* Wiley, New York.

5. Cooper, R.B. 1981. *Introduction to Queueing Theory.* North-Holland, New York.

6. Cox, D.R. and Smith, W.L. 1961. *Queues.* Wiley, New York.

7. Çınlar, E. 1975. *Introduction to Stochastic Processes.* Prentice- Hall, New Jersey.

8. Doshi, B.T. 1985. A Note on Stochastic Decomposition in a $GI/G/1$ Queue with Vacations or Set-up times. *J. Appl. Prob.* **22**, 419-428.

9. Doshi, B.T. 1986. Queueing Systems with Vacations-A Survey. *Queueing Syst. Theory Appl.* **1**, 29-66.

10. Eisen, M., and Tainiter, M. 1963. Stochastic Variations in Queueing Processes. *Opns. Res.* **11**, 922-927.

11. Federgruen, A., and Green, L. 1986. Queueing Systems with Service Interruptions. *Opns. Res.* **34**, 752-768.

12. Federgruen, A., and Green, L. 1988. Queueing Systems with Service Interruptions II. *Nav. Res. Log.* **35**, 345-358.

13. Fischer, M.J. 1977. An Approximation to Queueing Systems with Interruptions. *Mgmt. Sci.* **24**, 338-344.

14. Fuhrmann, S.W., and Cooper, R.B. 1985. Stochastic Decompositions in the $M/G/1$ Queue with Generilized Vacations. *Opns. Res.* **33**, 1117-1129.

15. Gaver, D.P. 1968. Diffusion Approximation and Models for Certain Congestion Problems. *J. Appl. Prob.* **5**, 607-623.

16. Gelenbe, E., and Pujolle, G. 1987. *Introduction to Queueing Networks*. Wiley, New York.

17. Gross, D., and Harris, C.M. 1974. *Fundamentals of Queueing Theory*. Wiley, New York.

18. Halachami, B., and Franta, W.R. 1978. A Diffusion Approximation to the Multi-Server Queue. *Mgmt. Sci.* **24**, 522-529.

19. Heyman, D.P., and Sobel, M.J. 1982. *Stochastic Models in Operations Research.* Vol.1. McGraw-Hill, New York.

20. Iglehart, D.L. 1965. Limit Diffusion Approximations for the Many- Server Queue and the Repairman Problem. *J. Appl. Prob.* **2**, 429-441.

21. Jaiswal, N.K. 1968. *Priority Queues*. Academic Press, New York.

22. Kendall, D.G. 1953. Stochastic Processes Occuring in the Theory of Queues and Their Analysis by the Method of the Embedded Markov Chain. *Ann. Math. Stat.* **34**, 338-354.

23. Kingman, J.F.C. 1965. The Heavy Traffic Approximation in the Theory of Queues. In *Proceedings of the Symposium on Congestion Theory* , E.L. Smith and W.E. Wilkerson (Eds.) , University of North Carolina Press, Chapel- Hill, North Carolina.

24. Kleinrock, L. 1975. *Queueing Systems: Theory*. Vol.1. Wiley, New York.

25. Kleinrock, L. 1976. *Queueing Systems: Computer Applications.* Vol 2. Wiley, New York.

26. Kobayashi, H. 1974. Application of the Diffusion Approximation to Queueing Networks: Parts I and II. *JACM* **21**, 316-328 and 459-469.

27. Miller, L.W. 1964. Alternating Priorities in Multi-Class Queues. Ph.D. dissertation. Cornell University.Ithaca, New York

28. Mitrany, I.L., and Avi-Ithzak, B. 1968. A Many Server Queue with Service Interruptions. *Opns.Res.* **16**, 628-638.

29. Neuts, M. 1981. *Matrix Geometric Solutions in Stochastic Models.* John Hopkins University Press, Baltimore, Maryland.

30. Neuts, M.F., and Lucantoni, D. 1979. A Markovian Queue with N Servers Subject to Breakdowns and Repairs. *Mgmt. Sci.* **25**, 849-861.

31. Newell, G.F. 1974. *Applications of Queueing Theory.* Springer- Verlag, New York.

32. Prabhu, N.U. 1987. A Bibliography of Books and Survey Papers on Queueing Systems: Theory and Applications. *Queueing Systems* **2**, 393-398.

33. Scholl, M., and Kleinrock, L. 1983. On the $M/G/1$ Queue with Rest Periods and Certain Service-Independent Queueing Disciplines. *Opns.Res.* **31**, 705-719.

34. Shanthikumar, J.G. 1988. On Stochastic Decomposition in $M/G/1$ Type Queues with Generalized Server Vacations. *Opns. Res.* **36**, 566-569.

35. Stoyan, D. 1977. Bounds and Approximations in Queueing through Monotonicity and Continuity. *Opns. Res.* **25**, 851-863.

36. Takács, L. 1962. *Introduction to the Theory of Queues.* Oxford University Press, New York.

37. Tejhen, J., Jr. 1986. Control of the Service Process in a Queueing System. *Eur. J. Opnl. Res* **23**, 141-158.

38. Thiruvengadam, K. 1963. Queueing with Breakdowns. *Opns. Res.* **11**, 62-71.

39. White, H., and Christie, L.S. 1958. Queueing with Preemptive Priorities or Breakdowns. *Opns. Res.* **6**, 76-96.

Chapter 4

SIMULATION OF QUEUEING SYSTEMS

M. Akif Eyler
Department of Industrial Engineering
Bilkent University
Ankara, Turkey

1. SIMULATION OF SIMPLE QUEUES

With the widespread availability and the ever-decreasing prices of computers, simulation has been gaining more and more popularity every year in many branches of technology. Simulation can be defined as "experimentation on computer." This phrase summarizes all the strengths and the weaknesses of simulation methodology. In order to illustrate this statement, let us outline how to simulate a simple queue in some detail.

In order to conduct a simulation study, the first step is the understanding of the system being studied. Suppose such an analysis indicates that the system can be modeled as a $G/G/1$ queue, with some arrival and service time distributions. These distributions may be summarized as a table of numbers or in the form of a mathematical function. (Obviously, only very limited classes of all possible distributions will have names in probability theory, such as normal or exponential distributions.) The determination of the probability distributions is inherently an emprical process and may require the use of some well-known statistical techniques.

The next step is the development of a computer program that simulates the behavior of the system. Notice that the program development should be considered "just one step" in the process of simulation study. Although a very crucial step, it is not by any means all of a simulation methodology. Such a program for the $G/G/1$ queue is presented later in this chapter.

Then comes the design of the experiments. The parameters that can be varied (usually within some constraint set) must be identified and the measures of performance must be selected. In the case of a $G/G/1$ queue, some of these parameters are:

T_a =Mean inter-arrival time $= 1/(\text{Arrival rate}) = 1/\lambda$
T_s =Mean service time $= 1/(\text{Service rate}) = 1/\mu$

Clearly, these are the only parameters that can be varied in the case of an $M/M/1$ queue, since an exponential distribution is completely described by its mean value. In the case of a $G/G/1$ queue, however, since more general distributions are allowed, there are many other parameters of interest, e.g., the variance of a normal distribution, or the upper and lower limits of a uniform distribution.

The performance measures, in general, are not related in a trivial way to these parameters. No simple formula exists to calculate these measures for $G/G/1$ queue. They constitue the output of the simulation. The following list contains some measures that are most often used in practice:

$$u = \text{Utilization of the server,}$$
$$W = \text{Mean time spent in the system,}$$
$$W_q = \text{Mean time spent in the queue (average waiting time),}$$
$$L = \text{Mean number in the system,}$$
$$L_q = \text{Mean number in the queue (average queue length).}$$

In fact, some useful identities between these measures make most of them redundant. Under very mild assumptions, the following properties can be shown to hold:

$$u = T_s/T_a = \lambda/\mu \tag{1}$$
$$L = \lambda W \text{ (Little's Law)} \tag{2}$$
$$L_q = \lambda W_q \tag{3}$$
$$W = W_q + T_s \tag{4}$$
$$L = L_q + u \tag{5}$$

Thus, it is not necessary to collect data for all of these measures. Accumulating the total time spent in the system for all customers will be sufficient to calculate the other measures. Let σ denote the total system time for all customers, T denote the observation period, and N denote the total number of customers served during that time. Then,

$$L = \sigma/T \qquad W = \sigma/N \tag{6}$$

will determine L and W. The corresponding quantities related to the queue can be obtained from (4) and (5). In fact, (2) can be proved by using (6) and the fact that the expected value of N/T is just λ.

At this point in our hypothetical simulation study, we are ready to execute the program for various values of the parameters and obtain the corresponding values for the measures of performance. With the computer technology of the last decade, this step was considered to be the most expensive one: it usually took a very long time on the precious computer. As computers become much more powerful and much less expensive, the relative importance of the actual execution of the simulation program is diminishing.

The final step in every simulation study is the analysis of the experimental results. This is where we need the help of the well- established field of statistics. The analysis of simulation output is similar in many ways to the analysis of experimental results in any other field of science, although there are a few more problems that have to be considered. These problems will be summarized in the subsequent sections.

Before discussing how to write a program for a $G/G/1$ queue, it may be good to describe a state transition model of this system. In order to characterize a $G/G/1$ queue completely, it is sufficient to know the arrival process and the service process. But since we cannot assume any probability distribution for these processes, the exact time of each arrival and the exact duration of each service must be known.

Input values:

N = Total number of customers served,
X_k = Time between the arrival of customers $k - 1$ and k $(k = 1, ..., N)$,
D_k = Duration of service for customer k $(k = 1, ..., N)$.

Now it is possible to calculate a set of variables based on these input values, using the recursive relations described below:

Derived values: (for $k = 1, ..., N$)

A_k = Arrival time for customer k,
S_k = Start of service for customer k,
E_k = End of service for customer k,
W_k = Waiting time for customer k.

The calculation of these values will be through the use of the equations below (for $k = 1, ..., N$):

$$A_k = A_{k-1} + X_k, \qquad A_0 = 0 \tag{7}$$

$$S_k = \max(A_k, E_{k-1}), \qquad E_0 = 0 \tag{8}$$

$$E_k = S_k + D_k \tag{9}$$

$$W_k = S_k - A_k . \tag{10}$$

Simulation of a $G/G/1$ queue could be defined as the iterative evaluation of these formulas on the computer. But there is a little flaw in this definition: although very well-defined sequences of numbers, they are just too many in number. Considering that N might well be hundreds of thousands, it may be impossible to store all of these numbers. Thus, a need arises for some summary values. It turns out that for this simplest queueing system a few such values will be sufficient for the calculation of the parameters and the performance measures defined above.

Various statistics of interest can be calculated as follows:

$$T = \text{Total time to serve } N \text{ customers} = E_N \tag{11}$$

$$\sigma = \text{Total system time for all customers} = \sum_k D_k + \sum_k W_k \tag{12}$$

$$T_a = T/N , \qquad T_s = \sum_k D_k/N \tag{13}$$

Now, (1)-(6) may be used to evaluate u, W, L, W_q, and L_q as desired. Table 1 contains, as illustration of these formulas, a very brief example of the application of (7)-(10) for $N = 10$.

Table 1. Hand Simulation of a $G/G/1$ Queue

k	X	D	A	S	E	W
1	5	4	5	5	9	0
2	3	4	8	9	13	1
3	16	4	24	24	28	0
4	5	4	29	29	33	0
5	1	4	30	33	37	3
6	2	4	32	37	41	5
7	3	4	35	41	45	6
8	17	4	52	52	56	0
9	2	4	54	56	60	2
10	5	4	59	60	64	1

The calculation of the statistics is shown below:

$$T = 64 \text{ min}$$
$$\sigma = 40 + 18 = 58 \text{ min}$$
$$T_a = 6.4 \text{ min}$$
$$T_s = 4.0 \text{ min}$$

The performance measures can also be calculated as easily:

$$u = 40/64 = 0.63$$
$$L = 58/64 = 0.91$$
$$L_q = 18/64 = 0.28$$
$$W = 58/10 = 5.8 \text{ min}$$
$$W_q = 18/10 = 1.8 \text{ min}$$

The information in Table 1 is presented in a graphical form in Figure 1. This graph shows the number of customers in the system as a function of time. Notice that σ is just the area under this curve. Moreover, L is the time average of the number of customers, and W is the time per customer (another average value).

A program that simulates a $G/G/1$ queue must do calculations equivalent to those in (7)-(10). Although it is possible to write a program that is identical to those equations, a more general approach is taken in Program 1. Only the state transitions are included in this program; statistics are not collected in order to show the basic skeleton of an event-driven simulation.

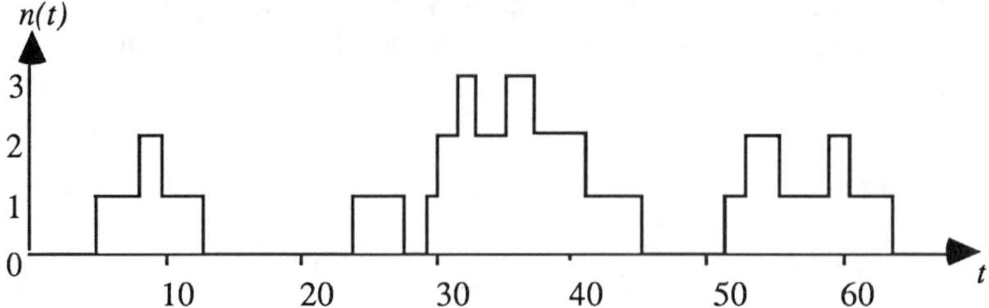

Figure 1. Plot of $n(t)$ in Table 1

The basic structure of this algorithm can be summarized as follows:

```
initialize;
repeat
    set t to next event time
    collect statistics (omitted in Prog.  1)
    perform event routine
until end of simulation
report results (omitted in Prog.  1)
```

The reason why this is is called the "Event-Oriented Method" should be obvious: time does not run in uniform steps, rather it only moves to the next event time. As nothing of interest happens between event times, this approach is acceptable to the modeler.

```
program G/G/1; (* Simulation of G/G/1 queue *)

const
  Inf= 1E30; (* A large number, stands for infinity *)
  Ta = 5; (* Mean inter-arrival time *)
  Da = 2; (* Maximum deviation of inter-arrival time *)
  Ts = 4; (* Mean service time *)
  Ds = 3; (* Maximum deviation of service time *)
  tF = 5000; (* Final time *)
var

  t:  Real; (* Time *)
  A: Real; (* Time of next arrival *)
  E: Real; (* Time of next service completion *)
  n:  Integer; (* Number in system *)

function Uni(T, D: real):  real;
(* generates a number from a uniform distribution with
   mean T and maximum deviation D, i.e., in [T-D,T+D] *)
   begin
```

```
      uni := T + D * (2*random-1)
   end;

 begin
    t:=0; A:=uni(Ta, Da); E:=Inf; n:=0;
    repeat
      if A<E then t:=A else t:=E;
      (* Event routines *)
      if t=E then begin (* Completion of service *)
        n:=n-1;
        if (n>0) then E:=t+uni(Ts, Ds)
          else E:=Inf
      end;
      if t=A then begin (* Arrival of customer *)
        n:=n+1; A:=t+uni(Ta, Da);
        if (n=1) then E:=t+uni(Ts, Ds)
      end
    until t>tF
 end.
```

Program 1. Simulation of a $G/G/1$ Queue

There are only two types of events in a simple queue: the arrival of a customer and the end of a service. The beginning of a service is not modeled as an event in this approach, because it always coincides with one or the other two types of events. Thus, the next event time is always the minumum of two numbers: the time of the next arrival or the end time of the current service, if any. Whether or not the server is currently busy can be seen by the inspection of either of two conditions: $E = \infty$ (i.e., the current service ends at infinity) or $n = 0$ (i.e., there are no customers in the system).

Obviously, the heart of the program lies in the event routines. The two routines for the simple queue are imbedded within an "if" statement. Whenever a customer arrives, n is incremented, time of next arrival is calculated, and a service starts if the server happens to be free at that moment. Whenever a service is completed, n is decremented, if there are no more customers the server becomes idle, otherwise another service starts. As a specific case of a $G/G/1$ queue, uniform distribution has been used for both the arrival and the service processes in Program 1. Clearly, other distributions can easily be substituted, if required.

Program 1 as described above correctly simulates the system under consideration, but it is quite useless as such: it fails to report any result of the simulation. In order to collect statistics, we can define three more variables:

```
var sum, prev, busy:  real;
```

which are initialized with `sum:=0; busy:=0;` and updated as follows:

```
prev:=t; if A<S then t:=A else t:=S;
```

```
sum := sum + n * (t-prev);
if n>0 then busy := busy + (t-prev);.
```

An alternative in the development of the program is the use of a model-oriented language. Figure 2 illustrates the use of one such language, GPSS. Designed basically for the simulation of queueing systems, this is a very efficient modeling tool.

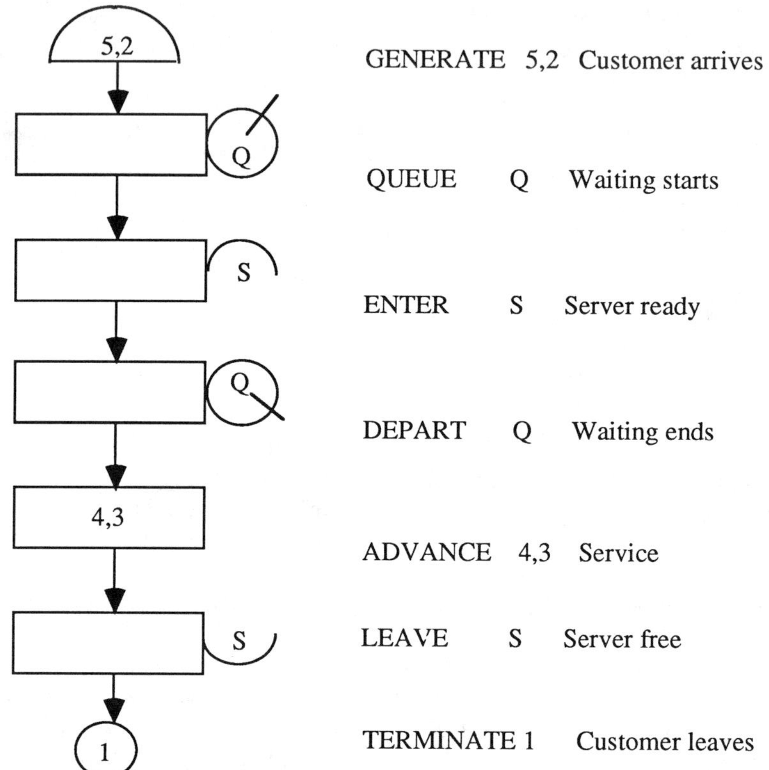

Figure 2. GPSS Program for a Simple Queue

GPSS and similar languages (e.g., SLAM, SIMAN) are based upon the use of a flow diagram. Such a diagram makes the communication easier between the analyst and the program. GPSS flow diagrams contain some 30 block types. Each block type has a specific meaning in the modeling process. Each of these blocks has a corresponding statement of the language. By combining only 7 of these statements in a proper way, it is possible to write a GPSS program that models a simple queue.

GPSS works around the concept of a "transaction", which corresponds to a customer in queueing theory. A transaction is created in a GENERATE block, and it moves as much as it can. There are mainly two types of reasons why it may not move:

1. it may enter an ADVANCE block, in which case it spends some time there, usually with a random duration from a given distribution.
2. it may need the attention of a server, indicated by an ENTER block, in which case it is stuck in the previous block, and it stays there until the condition is satisfied.

The fundamental difference between these two cases is that the duration of waiting is known (maybe statistically) in 1, but it depends on the state of the system in 2.

The other block types in Figure 2 are QUEUE-DEPART pair to collect statistics related to the queue, LEAVE to set the server free, and TERMINATE to kill the transaction.

2. SIMULATION OF QUEUEING NETWORKS

A simple queue may be extended in many ways: increase the number of servers, put a limit on the size of the waiting line, let the service rate depend on the queue length, let the arrival rate be an explicit function of time. Each of these extensions introduce so much difficulty into the analytical queueing model, that a previously solvable model may become computationally intractable as a result of this extension. A simulation model, on the other hand, is affected only very little by such extensions. In fact, all of these extensions could be incorporated into Program 1 within minutes by an experienced programmer. The interested reader is urged to try these modifications.

We would like to extend the simple queue of the previous section in a different direction: two simple queues in sequence. A customer arrives at the system and waits for the service as before. But, when this service is completed, it enters a second queue to wait for the attention of another server. The customer leaves the system on the completion of the second service. Clearly, the number of service stages can be increased as much as desired; however, it is very rare to see more than 2 or 3 stages in real applications.

A queueing system with two sequential stages is illustrated in Program 2. There are two kinds of servers named "Clerk" and "Keeper". The queues in front of these servers are labeled "One" and "Two", respectively. SEIZE and RELEASE blocks are very similar to ENTER and LEAVE blocks in function. The former pair is used when there is absolutely one server of that type, whereas the latter indicates the possibility of multiple servers. The STORAGE statement defines the number of such servers. This explanation should make clear that there are 2 Clerks and only 1 Keeper in Program 2.

Another extension of a simple queue toward queueing networks is the selection among multiple queues. In such a system, the customer may visit one of a given number of server classes. Each server class may have more than one server.

Program 3 is an example of a system with two types of servers, A and B. Here, parts arrive every 6 minutes to be processed by one of the two machines. They have a 90% probability of visiting the fast machine A. The rest of the parts go to the slow machine B. The branching is made possible by a TRANSFER block. As a further complication,

```
10              GENERATE     5,2
20              QUEUE        ONE
25              ENTER        CLERK
30              DEPART       ONE
35              ADVANCE      8,4
40              LEAVE        CLERK
50              QUEUE        TWO
55              SEIZE        KEEPER
60              DEPART       TWO
65              ADVANCE      4,3
70              RELEASE      KEEPER
90              TERMINATE    1
95    CLERK     STORAGE      2
```

Program 2. Two Servers in Sequence

both machines have some defective output. In this case 10% of the parts are rejected after either machine. As a simplification of the program, QUEUE-DEPART pairs are not shown. This simply means that the statistics will not be collected, but the queues still exist before the SEIZE blocks.

```
10              GENERATE     6
20              TRANSFER     0.9,    BB
30    AA        SEIZE        A
35              ADVANCE      4,2
40              RELEASE      A
45              TRANSFER     0.9,    REJECT
50              TERMINATE    1
60    REJECT    TERMINATE
70    BB        SEIZE        B
75              ADVANCE      20,10
80              RELEASE      B
90              TERMINATE
```

Program 3. Choose One of the Two Servers

A simplified model of a port is shown in Program 4. It is an example from Gordon (1978,p.221):

Ships arrive at a harbor at the rate of every 2 ± 1 hours. There are six berths to accomodate them. They also need the services of a crane for unloading and there are five cranes. After unloading, 10% of the ships stay to refuel before leaving; the others leave immediately. Ships do not need the cranes for refueling. It takes 15 ± 6 hours to unload and 2 ± 1 hours to refuel.

```
100            GENERATE    2,1
105            QUEUE       BQ
110            ENTER       BERTH
115            DEPART      BQ
120            QUEUE       CQ
130            ENTER       CRANE
140            DEPART      CQ
150            ADVANCE     15,6
155            LEAVE       CRANE
160            TRANSFER    .9,,ZAP
180            ADVANCE     2,1
190    ZAP     LEAVE       BERTH
200            TERMINATE   1
210    *
250    BERTH   STORAGE     6
260    CRANE   STORAGE     5
```

Program 4. Simulation of a Port

The solution to this problem is given using only the blocks that were introduced before. In some way, this example constitutes a combination of the previous two: there are two services in sequence, but the second is optional. There is an additional server (berth) that is with the transaction as long as it is in the system.

These examples should be taken as very special cases of queueing networks. A more general network is made up of servers with the respective queues connected in some logical way. But the simulation of such systems does not require any other concept than those considered so far. Although the analysis may become much more difficult with the addition of another node, the simulation itself may be conducted using the same program.

3.   COMPUTATIONAL PROBLEMS IN SIMULATION

When the system being simulated is more complicated than a simple queue, some computational problems related to simulation methodology must be expected. One of these problems is the determination of the next event. When there are only two types of events, the determination of the next event is just the minimum of two numbers. However, when there are more event types, an event list must be maintained. Such a list will contain records for each future event, indicating the event type and the event time. Event lists are usually ordered in time. This simplifies the event selection problem by an order of magnitude.

As the number of event types increase, the modeling process may become more difficult. An approach to reduce the conceptual complexity of the model is the introduction of C-events, the usual events being called B-events. B and C in these names stand for

"Bound" and "Conditional", respectively. B-events, like the arrival of a customer or the end of a service, are determined by other events that occur before. The occurrence of C-events depend on the state of the system, like the start of a service in a simple queue. This method is called "The three-phase approach", in contrast to the two-phase approach discussed earlier. The basic structure of the three-phase algorithm can be summarized as follows:

```
initialize;
repeat
    set t to next event time
    collect statistics (omitted in Prog.  5)
    perform B-events
    perform C-events
until end of simulation
report results (omitted in Prog.  5)
```

This structure is detailed in Program 5, where the declarations are also omitted as they are the same as in Program 1. Although this approach has a slightly longer code and it may take longer to execute, it is highly recommended because of the reduction it may introduce at a conceptual level. Thus, Program 5 may be easier to develop and to follow than Program 1.

```
begin
    t:=0; A:=uni(Ta, Da); E:=Inf; n:=0;
    repeat
      (* Set clock *)
      if A<E then t:=A else t:=E;
      (* B-Events:  *)
      if t=E then begin (* Completion of service *)
         n:=n-1; E:=Inf
      end;
      if t=A then begin (* Arrival of customer *)
         n:=n+1; A:=t+uni(Ta, Da)
      end;
      (* C-Events:  Start service *)
      if (E=Inf) and (n>0) then E:=t+uni(Ts, Ds)
    until t>tF
end.
```

Program 5. Three-Phase Approach

4. REFERENCES

Two early textbooks, one at an introductory level by Gordon (1978), the other at a more advanced level by Fishman (1978) are still useful today. Law and Kelton (1982) present a treatment for statistical problems in simulation. Banks and Carson (1984) include a detailed chapter on queueing models, rich with many examples. Solomon (1983) gives many examples on the simulation of queueing systems.

A monograph by Rubinstein (1986) deals exclusively with some theoretical issues related to the simulation of queueing networks. A chapter on Perturbation Analysis is included. MacDougall (1987) gives interesting examples from the simulation of computer systems. A recent overview by Glynn and Iglehart (1988) summarizes simulation methods in queues.

A speculative paper by Ho (1987) is recommended for a readable overview of Perturbation Analysis from the perspective of control theory. This paper includes many references and research ideas for anyone interested in the topic.

1. Banks,J. and Carson,J.S. 1984. *Discrete-Event System Simulation.* Prentice-Hall, Englewood Cliffs, New Jersey.

2. Fishman,G.S. 1978. *Principles of Discrete Event Simulation.* Wiley, New York.

3. Glynn,P.W. and Iglehart,D.L. 1988. Simulation Methods for Queues: an Overview. *Queueing Systems* **3**, 221-256.

4. Gordon,G. 1978. *System Simulation.* Prentice-Hall, Englewood Cliffs, New Jersey.

5. Ho,Y.C. 1987. Performance Evaluation and Perturbation Analysis of Discrete Event Dynamic Systems. *IEEE Trans. on Auto. Cont.* **AC-32**, 563-572.

6. Law,A.M. and Kelton,W.D. 1982. *Simulation Modeling and Analysis.* Mc-Graw-Hill, New York.

7. MacDougall,M.H. 1987. *Simulating Computer Systems.* MIT Press, Cambridge, Massachusetts.

8. Rubinstein, R.Y. 1986. *Monte-Carlo Optimization, Simulation and Sensitivity of Queueing Networks.* Wiley, New York.

9. Solomon,S.L. 1983. *Simulation of Waiting-Line Systems.* Prentice-Hall, Englewood Cliffs, New Jersey.

Chapter 5

QUEUEING NETWORKS:
A SURVEY OF
ANALYTICAL RESULTS

Ali Rıza Kaylan
Department of Industrial Engineering
Boğaziçi University
Bebek, Istanbul, Turkey

1. INTRODUCTION

Queueing network models are frequently encountered in diverse application areas such as telecommunication, production, computer, transportation, health care and other service systems. Following the fundamental works of Erlang (1917), the first key paper on this topic is of Jackson (1957) which is titled "Networks of Waiting Lines". The interest in these models has grown substantially in the past three decades and a comprehensive survey carried out by Disney and König (1985) reports more than 300 related articles. This chapter attempts to provide an overview of available analytical results. It is intended to emphasize the concepts and tools of analysis such as reversibility, conservation laws, insensitivity, and Coxian distributions.

A queueing network is simply depicted as a set relation $< \mathcal{K}, \mathcal{A} >$ where $\mathcal{K}$ is a collection of K nodes and A is the set of arcs connecting these nodes. Each node represents a service system composed of a service mechanism and its associated waiting line. Customer flow is realized on the connecting arcs without any delays. A switching process governs the traffic among these nodes. Let $P = [P_{ij}]$ denote the $K \times K$ matrix containing these transition probabilities. Apart from this internal traffic, let $\lambda_0 = (\lambda_{01}, \lambda_{02}, ..., \lambda_{0K})$ be the row vector of external arrival rates. Then, the total arrival rate including exogenous arrivals as well as the internal transfers coming from the other nodes can be represented as a row vector $\lambda = (\lambda_1, \lambda_2, ..., \lambda_K)$ where

$$\lambda = \lambda_0 + \lambda P$$

with individual elements

$$\lambda_j = \lambda_{0j} + \sum_{i=1}^{K} \lambda_i P_{ij}, \qquad j = 1, 2, ..., K$$

showing the equilibrium rate of flow through node j. Figure 1 describes the structure of node j in equilibrium. In order to study the system behaviour, a full characterization of the system needs to be specified. Namely, the number of servers $m = (m_1, m_2, ..., m_K)$ at each node and their associated service distributions with corresponding speeds $\mu_j(i)$, $i = 1, 2, ..., m_j$, the waiting room capacities and the scheduling rules must be also provided.

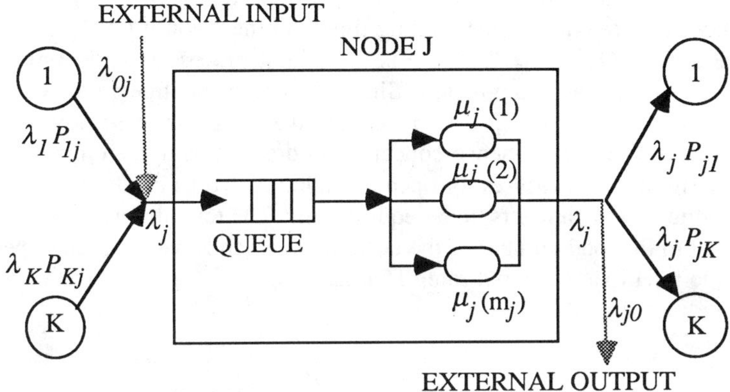

Figure 1. Structure of Node j in Equilibrium

There are several stochastic processes which are of interest in queueing networks. The number of customers in the queueing network is represented by a vector-valued random process $N(t) = (N_1(t), N_2(t), ..., N_K(t)), t \geq 0$, where $N_j(t)$, $j = 1, 2..., K$, is the number of customers at node j at time t, and it includes those in the waiting line plus those in service. This system size process which has discrete state space and continuous time parameter can also be studied at discrete time points, namely the arrival or departure epochs of customers resulting in the embedded queue length processes. The analysts are concerned with the form of the distribution function in steady-state, $p(n_1, n_2, ..., n_K)$ and whether the product form solution exists. The product form solution is simply expressed as the joint probability function being equal to the product of the marginal probability functions. That is,

$$p(n_1, n_2, ..., n_K) = \prod_{j=1}^{K} p_j(n_j)$$

which means the number of customers at each node is independent of the queue length at other nodes. The presence or absence of the insensitivity property, e.g., results being only dependent on the mean service time and not on the service time distribution, is another issue of interest.

The total waiting time of customer n is defined as

$$W_n = W_{n1} + W_{n2} + \cdots + W_{nK}$$

where W_{nj} is the time spent in the queue at node j by customer n. Thus, the sequence $\{W_n, \ n = 1, 2, ...\}$ is the waiting time process. The sojourn time of nth arriving customer S_n is the total system time which elapses from entry to the network until exit. The sequence $\{S_n, \ n = 1, 2, ...\}$ is the sojourn time process. Note that

$$S_n = S_{n1} + S_{n2} + \cdots + S_{nK}$$

where S_{nj} is the total system time at node j for customer n. Very few results are available about the structure of the waiting time and the sojourn time processes.

The sequence $\{O_{mj}, m = 1, 2, ...\}$ is the output process at node j where O_{mj} is the time of the mth service completion. Similarly, one can define an arrival process for each node. An interesting issue to be investigated is whether these flow processes are reversible or not. The independence concept also draws attention from researchers.

Many performance measures for open queueing networks can be expressed in terms of just the traffic flow parameters under equilibrium. The total throughput of the network which is equal to the total number of the customers which leave the system per unit time is equal to the total external input rate. That is,

$$\lambda = \sum_{j=1}^{K} \lambda_{0j}$$

which is also equal to

$$\lambda = \sum_{j=1}^{K} \lambda_{j0}.$$

where $\lambda_{j0} = \lambda_j (1 - \sum_k P_{jk})$.

Average use of the service facility at node j is measured by the traffic intensity $\rho_j = \lambda_j / \mu_j$ where $\mu_j = \sum_i \mu_j(i)$ is the overall service rate at node j. If there is only one server at node j, then the server utilization factor ρ_j is equal to the fraction of the time the server is busy. The total traffic intensity

$$\rho = \rho_1 + \rho_2 + ... + \rho_K$$

designates the total average number of customers being served in the network. Dividing this by λ yields

$$E[S] = \rho / \lambda$$

which is the total average amount of service received by a customer in the network.

The "visit ratio" which is the average number of requests per customer from node j can easily be obtained from the traffic balance equations. This is equal to

$$e_j = \lambda_j / \lambda, \qquad j = 1, 2, ..., K$$

which indicates the average number of visits to node j by a customer.

If the steady-state distribution function for the system size is obtained, then the average number of customers at node j is

$$E[N_j] = \sum_{n=0}^{\infty} n p_j(n)$$

and the total average number of customers in the network is

$$E[N] = \sum_{j=1}^{K} E[N_j].$$

According to Little's theorem, one can obtain the average response time which elapses from the time customer steps into the network to its departure time as

$$W = E[N]/\lambda.$$

Similarly, the total waiting time per customer can be obtained as

$$W_q = W - E[S].$$

2. CLASSIFICATION OF QUEUEING NETWORKS

Queueing network models can be categorized into two classes in terms of its switching process P. The classification is reflected in Figure 2.

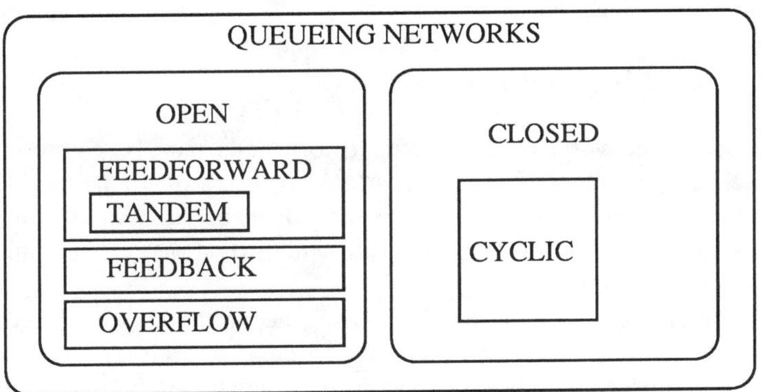

Figure 2. Network Classification According to its Switching Process

It is defined as a closed network if no exogenous arrivals or departures occur. That is,

$$\lambda_{0j} = \lambda_{j0} = 0$$

for all values of j. A fixed number of customers circulate in a closed network with no arcs moving the customers to the outside world. The row sums of P are equal to 1 (i.e., $\sum_j P_{ij} = 1$, $i = 1, 2, ..., K$). Otherwise, the network is open, and it has external input and output. In the open class, the network is referred as feedforward if the traffic flow is unidirectional. Feedforward networks are those networks in which customers are allowed to visit a node at most once. The nodes can be sequenced in such a way that the matrix P is upper triangular. A special subclass of feedforward networks are called as tandem networks. Their topology is simply a series of queues. In such sequential networks, all arrivals occur at the first node and departures take place at the last node. The switching process is defined as $P_{ij} = 1$ if $j = i + 1$, $i = 1, 2, ..., K - 1$. All other P_{ij}'s are equal to zero.

There is another special kind of open networks referred as overflow networks. This time, the switching of the arriving customers takes place in a sequential manner conditional on the status of the parallel nodes. K nodes are connected in parallel. Each arrival attempt to enter the first node. If there is no room, then it tries the others in the given order. If all nodes are full, then overflow takes place.

For closed networks, a subclass is defined as cyclic which is very similar to tandem networks except that there is no exogenous arrivals. Node K is connected back to node 1. That is, $P_{ij} = 1$ if $j = i + 1$, $i = 1, 2, ..., K - 1$, and $P_{K1} = 1$. All other transition probabilities are equal to zero.

Queueing networks are also classified according to the service time distribution either being exponential or general service times . All customers may belong to the same class with the same switching process and requesting similar services, or there may be different types of customers. In this way, single class or multiclass networks are obtained.

3. CUSTOMER FLOW PROCESSES

In a queueing network, the input stream to a node may be composed of several streams which are generated from the departure processes of the neighboring nodes as seen in Fgure 1. The first remarkable result we will focus on is due to Burke(1956).

Burke's Result: For an $M/M/m$ system with arrival rate λ, the output process is also Poisson with the same rate λ.

In other words, customer arrivals occurring according to a Poisson process and going through an exponential server with rate m generate Poisson departures with the same arrival rate independent of the service rate parameter μ. In order to illustrate this idea on an $M/M/1$ system, let the interdeparture probability density function be denoted as d(t) and its Laplace transform as $D^*(s)$. Recall that a departing customer will leave an empty system for the $M/M/1$ case with probability $1 - \rho$ where $\rho = \lambda/\mu$. In this case, the next interdeparture time will be the convolution of an interarrival time and a service time, both being exponential. If there is a waiting customer, then the interdeparture time will be just an exponential service time. The two possible cases are portrayed in Figure 3.

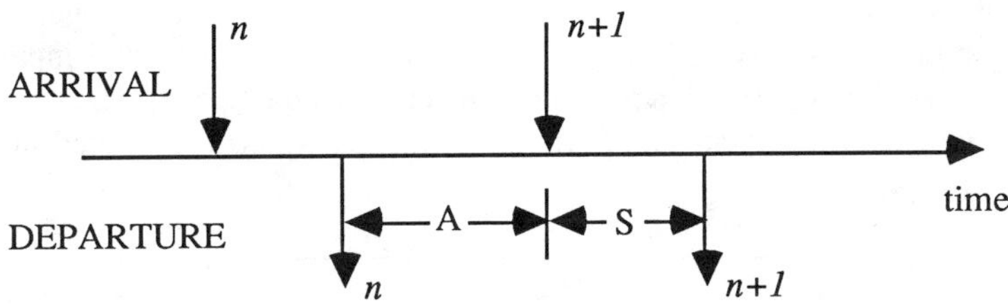

Case 1. System is left empty at the nth departure time

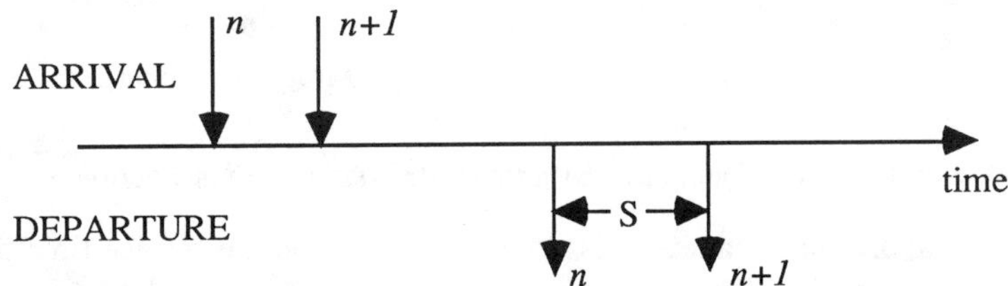

Case 2. System is left nonempty at the nth departure time

Figure 3. Interdeparture Time in Terms of Arrival and Service Times

The Laplace transform corresponding to d is

$$D^*(s) = (1 - \rho)A^*(s)B^*(s) + \rho B^*(s)$$

where

$$A^*(s) = \lambda/(\lambda + s)$$
$$B^*(s) = \mu/(\mu + s)$$

are the Laplace transforms for the exponential interarrival and service time distributions respectively. A straightforward algebraic manipulation yields

$$D^*(s) = \lambda/(\lambda + s)$$

which is equivalent to saying that the interdeparture distribution is exponential with rate λ. In fact, it can be shown that the $M/M/m$ is the unique FCFS queueing system with this property.

Suppose now that the independent Poisson processes coming out of various nodes are split according to a certain switching process and transferred to a neighboring node. Also these transfer streams are merged with an external arrival stream which is governed by another independent Poisson process. After splitting and merging, the total input stream to this neighboring node follows a Poisson process. This can be easily verified by recalling the fact that the decomposition and superposition of independent Poisson processes constitute a Poisson process.

It should be noted ,however, that under the assumptions of exogenous Poisson arrivals and exponential servers in an open queueing network, the total input stream to a node is not necessarily Poisson. To demonstrate this fact, consider a single server queueing system with instantaneous Bernoulli feedback which is reflected in Figure 4.

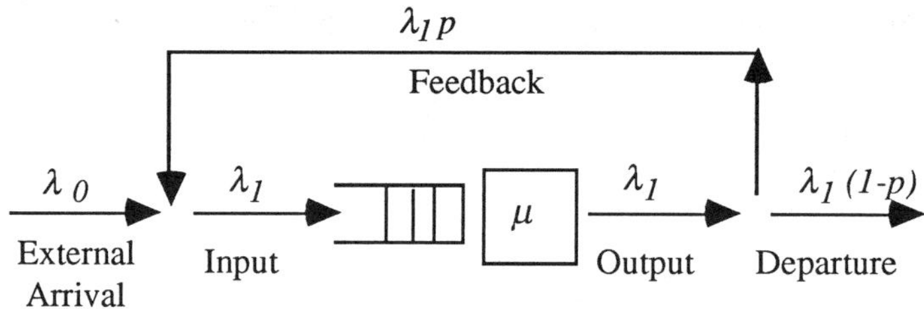

Figure 4. A Single Server Queueing System with Instantaneous Feedback

The external arrival stream is Poisson with rate λ_0 and the service time distribution is exponential with rate μ. The customer whose service is completed leaves the system with probability $1-p$ or instantaneously rejoins the queue with probability p. The traffic balance equation

$$\lambda_1 = \lambda_0 + \lambda_1 p$$

yields

$$\lambda_1 = \frac{\lambda_0}{(1-p)}.$$

It is interesting to note that the system size probability $p(n)$ under the steady-state condition $(\lambda_1 < \mu)$ is equivalent to the results of an $M/M/1$ system, but the interarrival time T which elapses between consecutive input instants (realized as external arrivals or feedbacks) is not an exponential distribution. By equating the instantaneous transition rates in the birth and death equations, the system of equations

$$p(0)\lambda_0 = p(1)\mu(1-p)$$
$$p(n)[\lambda_0 + \mu(1-p)] = p(n-1)\lambda_0 + p(n+1)\mu(1-p), \quad n = 1, 2, \ldots$$
$$\sum_{n=0}^{\infty} p(n) = 1$$

reveal the same solution as the $M/M/1$ system. That is,

$$p(n) = \rho^n(1 - \rho), \qquad n = 0, 1, 2, \ldots$$

where $\rho = \lambda_1/\mu$.

Now, denote the time to the next arrival to the queue as T at an arrival epoch. This interarrival time is

$$T = min(A, B)$$

where A and B are the times until the next exogenous arrival and the next feedback customer respectively. The distribution function of T is

$$P(T > t) = P(A > t)P(B > t)$$

since A and B are independent. Due to the memoryless property, it is known that

$$P(A > t) = exp\left[-\lambda_0 t\right].$$

The probability density function $f(t)$ of B can be obtained by conditioning on the system size. Suppose that the customer has found n in the system upon arrival. Thus, the system size would be n+1 including the customer. The probability density function of B is

$$f(t) = \sum_{n=0}^{\infty} p(n)\left[\sum_{j=1}^{n+1} p(1 - p)^{j-1} g_j(t)\right]$$

where $g_j(t)$ is the Erlang density with parameters j and μ. The terms inside the inner summation indicates that the jth customer among the $n + 1$ customers is the first to be fed back. Essentially, j exponential service times are convolved to yield an Erlang-type j distribution. Substituting the Erlang density function, it can be derived that

$$F(t) = p\mu[1 - exp\{-(\mu - \lambda_0)t\}]/(\mu - \lambda_0)$$

Thus, the distribution function for T is

$$P(T > t) = \{[(1 - p)\mu - \lambda_0]exp(-\lambda_0 t) + p\mu\ exp(-\mu t)\}/(\mu - \lambda_0)$$

which verifies our claim that the total input stream is not Poisson. The distribution of T is recognized to be a mixture of two exponential distributions known as a hyperexponential distribution of order 2

$$P(T > t) = p_1 exp(-\lambda_0 t) + (1 - p_1)exp(-\mu t)$$

where p_1 is the mixing probability given by

$$p_1 = \frac{(1-p)\mu - \lambda_0}{\mu - \lambda_0} \,.$$

It is quite extraordinary to find out that the stationary distribution of the system size for this feedback system is the same as that of an $M/M/1$ model. It is even more remarkable to find out in the following sections about Jackson networks that the queue lengths are independent of each other so that the product form solution exists. This is in fact true in spite of the fact that the customer flows on the instantaneous feedback arcs are not independent of those departing the given nodes.

4. REVERSIBLE PROCESSES AND BALANCE EQUATIONS

The output process for an $M/M/m$ system can also be illustrated by the notion of time reversibility. During a time period $[0, T]$, what is observed in the reverse time is stochastically equivalent to the forward time process in steady state. That is, the departure process in the forward time is viewed to be the arrival process in the reverse time.

Typical sample paths are traced in Figure 5 for a queueing system to illustrate the forward and reverse systems. The time points designated as a and d are the arrival and departure epochs respectively. The shaded areas between the step functions denoting the arrival and departure processes reflect the total time spent in the system by all customers. When $a(t) = d(t)$, the system empties itself. For a reversible process, what is observed in forward and backward times are stochastically equivalent.

Before saying more on the departure process, the reversibility method as introduced by Reich (1957) deserves more explicit treatment as an analysis tool.

Reversibility Definition: A stochastic process $\{N(t), t \geq 0\}$ is reversible if

$$P\{N(t_1) = i_1, N(t_2) = i_2, ..., N(t_n) = i_n\}$$
$$= P\{N(T - t_1) = i_1, N(T - t_2) = i_2, ..., N(T - t_n) = i_n\}$$

for any sequence of time points $t_1, t_2, ..., t_n$ over an arbitrary period $[0, T]$.

If we trace the sequence of states $m+1, m, m-1, ...$ backward in time in Markovian queueing models, the backward transition probabilities can easily be shown to satisfy

$$Q_{ij} = P\{N_m = j | N_{m+1} = i\} = \frac{p_j P_{ji}}{p_i} \quad i, j \geq 0.$$

To be able to say that the process is time reversible, Q_{ij}'s and P_{ij}'s are supposed to be the same. In any stationary Markov chain, the necessary and sufficient condition for this to hold true is

$$p_i P_{ij} = p_j P_{ji} \qquad i, j \geq 0$$

which is referred as the detailed balance equations.

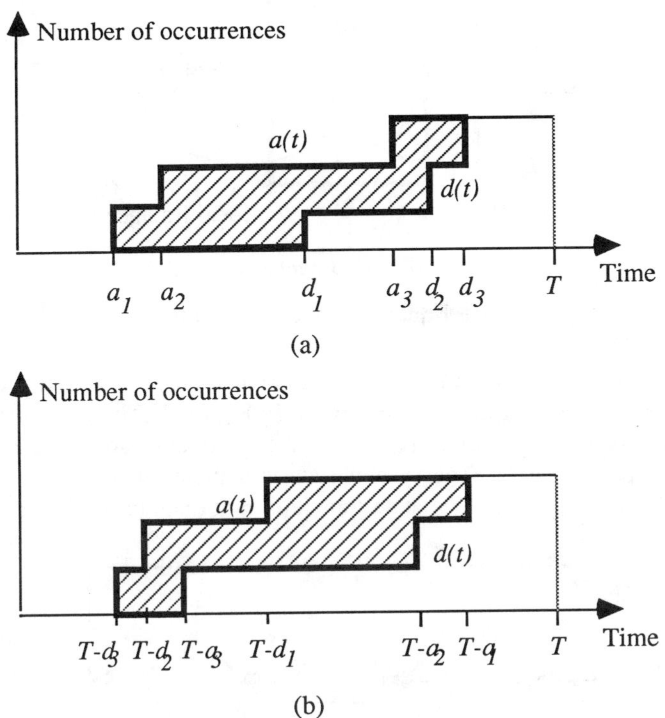

Figure 5. Sample Paths of Forward and Reverse Systems for the Time Period $[0, T]$

There are two other types of identities intimately related to flow conservation. These are called global and local balance equations. Global balance equations simply state that the input flow must equal to the output flow from a given state n. Local balance equations state that the rate of transition out of state n due to a customer departure from node i equal to the rate of transition into state n due to a customer arrival into node i.

To exemplify these equilibrium relations, consider the state transition rate diagram of a birth and death process in Figure 6. The global balance equations are obtained as

$$\lambda_{n-1}p_{n-1} + \mu_{n+1}p_{n+1} = (\lambda_n + \mu_n)p_n.$$

The local balance equations are constructed by simply drawing a vertical line between adjacent states and equating flows across those boundaries. Thus, the global balance equation is split into two local balance equations as

$$\lambda_{n-1}p_{n-1} = \mu_n p_n$$
$$\mu_{n+1}p_{n+1} = \lambda_n p_n.$$

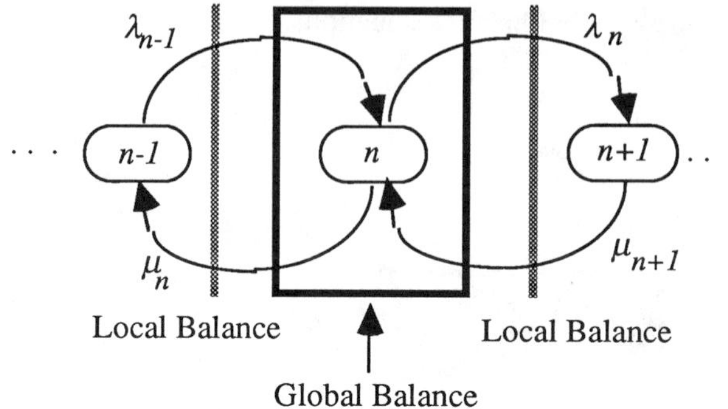

Figure 6. Illustration of Balance Equations Under Equilibrium

To elaborate further on the flow conservation equations, consider a cyclic queueing network composed of $K = 3$ nodes as portrayed in Figure 7. The customers circulating in this closed network is taken to be $N = 2$. The service times are exponential with rate μ_j at the jth respective node. The system state is depicted with a triplet (n_1, n_2, n_3), and there are 6 possible states. The state transition rate diagram is shown in Figure 8.

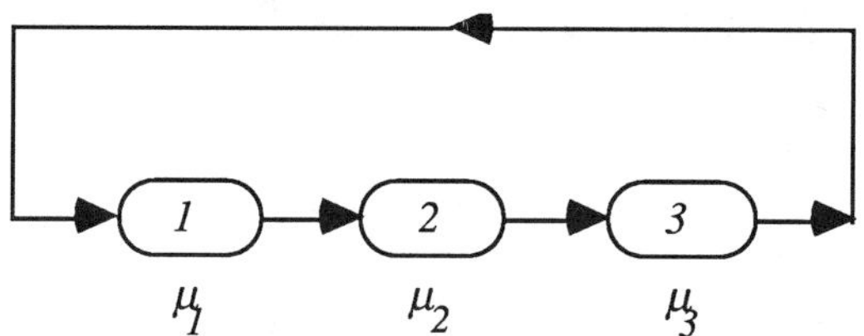

Figure 7. A Cyclic Queueing Network with Three Nodes and 2 Customers

The global balance equations for this network are obtained by equating the flow out of a state to the flow into that state. These equations are explicitly written as

$$\mu_2 p(0, 2, 0) = \mu_1 p(1, 1, 0)$$
$$\mu_1 p(2, 0, 0) = \mu_3 p(1, 0, 1)$$
$$\mu_3 p(0, 0, 2) = \mu_2 p(0, 1, 1)$$
$$(\mu_1 + \mu_2) p(1, 1, 0) = \mu_3 p(0, 1, 1) + \mu_1 p(2, 0, 0)$$
$$(\mu_2 + \mu_3) p(0, 1, 1) = \mu_1 p(1, 0, 1) + \mu_2 p(0, 2, 0)$$
$$(\mu_1 + \mu_3) p(1, 0, 1) = \mu_2 p(1, 1, 0) + \mu_3 p(0, 0, 2)$$

The first three of these equations are already in local balance form. Each of the last three can be split into two local balance equations. The dotted lines in Figure 8 shows which components are supposed to be in the same local balance equation. The detailed

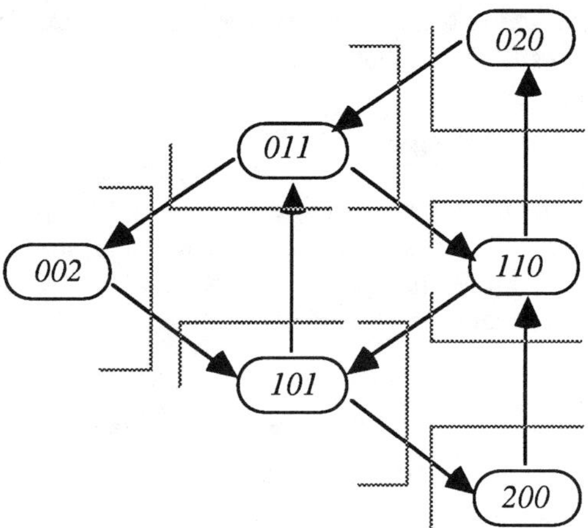

Figure 8. State Transition Rate Diagram for the Cyclic Network Example(The dotted lines indicate local balance.)

balance equations are obviously not satisfied for this example. Thus, the network is not a time reversible network.

Consider now a feedforward queueing network in which the external arrivals occur according to independent homogeneous Poisson processes and service facilities are exponential. There are m_j independent identical servers each serving with rate μ_j at node j, $j = 1, 2, ..., K$. The routing of customers takes place according to the switching matrix P which is upper triangular as defined in Section 2. In obtaining the steady-state system size probabilities, the first node is not influenced by the remaining nodes and it can be treated as separate from the others as an $M/M/m$ system. When the ergodicity condition $\lambda_1 < m_1\mu_1$ holds, the solution for the system size probability at node 1 is

$$p_1(n_1) = \frac{\lambda_1^{n_1} p_1(0)}{\prod_{j=1}^{n_1} \mu_1(j)}$$

where

$$p_1(0) = \Big[\sum_{n_1=0}^{\infty} \big(\lambda_1^{n_1} / \prod_{j=1}^{n_1} \mu_1(j)\big)\Big]^{-1}$$

and

$$\mu_1(j) = min(j, m_1), \qquad j = 1, 2, ... \; .$$

If we now focus on the arrival stream to node 2, it is already observed from Burke's result that it is also Poisson with rate

$$\lambda_2 = \lambda_{02} + \lambda_1 P_{12}.$$

Furthermore, at any fixed time t, the past departure stream from node 1 is independent of the current number of customers at that node. It is easier to observe this independence in reverse system. The past departures in forward system are the future arrivals in reverse time which obviously does not affect the current system size. This result can easily be extended to the remaining nodes.

We can conclude that the input processes to each node in a feedforward network are independent homogeneous Poisson processes and the marginal probabilities of system size can be obtained using the classical results of an $M/M/m$ queue. The stationary distribution of the network state is

$$p(n_1, n_2, ..., n_K) = \prod_{j=1}^{K} p_j(n_j)$$

where $p_j(n_j)$'s are the results of $M/M/m$ system provided that each $\lambda_j < m_j \mu_j$, $j = 1, 2, ..., K$.

5. JACKSON NETWORKS

Consider a totally open network of K nodes. This means any customer who visits any node of the network is certain to leave the network eventually. Exogenous arrivals occur according to independent, homogeneous Poisson processes with rates λ_{0i}, $i = 1, 2, ..., K$ to the ith node. The service mechanism at node i has m_i identical exponential servers with mean service time $1/\mu_i$ each. Once served at node i, a customer instantaneously goes to node j with probability P_{ij} or leaves the network with probability P_{i0}. The queueing discipline is FIFO, and there is no limitation on the capacities of the queues. The flow balance equations

$$\lambda_j = \lambda_{0j} + \sum_i \lambda_i P_{ij}, \qquad j = 1, 2, ..., K$$

have a unique solution since its matrix is invertible.

Jackson's Theorem: If the unique solution to the flow balance equations satisfies the conditions

$$\rho_i = \left[\frac{\lambda_i}{m_i \mu_i}\right] < 1, \qquad i = 1, 2, ..., K$$

then the steady-state distribution is obtained as a product form solution with individual marginal probabilities from $M/M/m$ queueing models. That is,

$$p(n_1, n_2, ..., n_K) = p(n_1)p(n_2)...p(n_K).$$

For the proof, recall the fact that for an irreducible Markov process if the balance equations yield a positive solution which also satisfies the normalization equation

$\sum_n p(n) = 1$, then this solution is the steady-state solution. Let us first write the global balance equations for the equilibrium probability distribution.

$$p(n_1, n_2, ..., n_K)\left[\sum_{j=1}^{K}\lambda_{0j} + \sum_{j=1}^{K}\mu_j(n_j)I_{(n_j>0)}(1 - P_{jj})\right]$$

$$= \sum_{j=1}^{K} p(n_1, ..., n_j - 1, ..., n_K)I_{(n_j>0)}\lambda_{0j}$$

$$+ \sum_{j=1}^{K} p(n_1, ..., n_j + 1, ..., n_K)\mu_j(n_j + 1)P_{j0}$$

$$+ \sum_{j=1}^{K}\sum_{\substack{i=1 \\ i \neq j}}^{K} p(n_1, ..., n_i + 1, ..., n_j - 1, ..., n_K)\mu_i(n_i + 1)I_{(n_j>0)}P_{ij}$$

for all $(n_1, n_2, ..., n_K) \geq 0$. Here, $\mu_j(k) = \mu_j \, min(k, m_j)$ for all $j = 1, 2, ..., K$ and $k \geq 0$.

The three components on the right-hand side represent the instantaneous transition rate into state n. These are the rates due to external arrivals, external departures, and the internal transfers in the given order. The left hand-side is the flow rate out of state n. By substituting the product form solution directly into the balance equation and checking the normalization equation, the theorem can easily be verified.

Gordon and Newell (1967) dealt with a closed network with N jobs circulating inside. Since all nodes communicate with each other, the corresponding Markov chain is irreducible. Its steady-state is given in the next theorem.

Gordon and Newell's Theorem: The steady-state distribution of a closed network is given by

$$p(n_1, n_2, ..., n_K) = \frac{1}{G(N)}\alpha_1(n_1)\alpha_2(n_2)...\alpha_K(n_K)$$

for all $n_j \geq 0$, $n_1 + n_2 + \cdots + n_K = N$ where

$$\alpha_i(n_i) = \frac{\lambda^{n_i}}{\prod_{j=1}^{n_i}\mu_i \, min(j, m_i)}$$

with λ_j as any nonzero solution to the balance equation

$$\lambda_j = \sum_{i=1}^{K}\lambda_i P_{ij}$$

and the normalization factor

$$G(N) = \sum_{n_1+n_2+\cdots+n_K=N}\alpha_1(n_1)\alpha_2(n_2)...\alpha_K(n_K)$$

The theorem can also be proved by direct substitution into the global balance equations and observing that they are satisfied. Note that the node states are dependent on each other in this case since the total number of circulating jobs is fixed.

6. COXIAN DISTRIBUTIONS

To relax the exponential distribution assumption in the Markovian queueing network models, one can employ a family of service distributions generalized by Cox (1955). The exponential, hyperexponential, and Erlang distributions are all in this Coxian family. Furthermore, any distribution function can be approximated by such a distribution.

The Laplace transformations for the hyperexponential distribution of order k and the Erlang-k distribution representing k servers in parallel and in series respectively are

$$F_1^*(s) = \sum_{i=1}^{K} \left[\frac{\alpha_i \mu_i}{\mu_i + s} \right]$$

$$F_2^*(s) = \prod_{i=1}^{K} \left[\frac{\mu_i}{\mu_i + s} \right] .$$

The service mechanism at a node is essentially treated as a group of service phases which are arranged either in series or parallel. To be able to start service on the next customer, all of the phases of the facility must be free.

It can be easily verified that the squared coefficient of variation

$$c^2 = \frac{\sigma^2}{E^2[X]}$$

is less than 1 for Erlang distributions and larger than 1 for hyperexponential distributions. Thus, series arrangements can reduce the variability of the service distribution, whereas parallel arrangements tend to increase it. In order to repesent the series-parallel structure of service mechanism with exponential service at each phase, Cox proposed a service facility given in Figure 9.

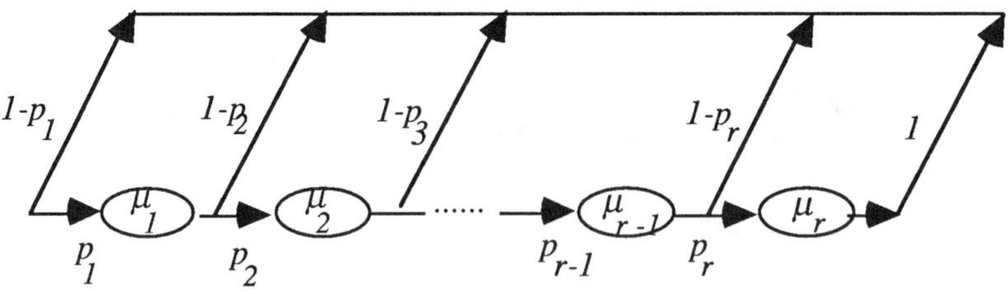

Figure 9. Coxian Phase-Type Distribution

The system is composed of r phases. Upon completing service at phase i, the customer moves to phase $i + 1$ with probability p_{i+1} for more service or leaves the facility with probability $1 - p_{i+1}$. The Laplace transform of this generalized phase-type distribution is

$$B^*(s) = (1 - p_1) + \sum_{i=1}^{r} p_1 p_2 ... p_i (1 - p_{i+1}) \prod_{j=1}^{i} \left[\frac{\mu_j}{\mu_j + s} \right]$$

where p_{r+1} is equal to 1. Clearly, $B^*(s)$ is a rational function of s which can be written as a ratio of two polynomials $P(s)/Q(s)$. Thus, any distribution with its Laplace transform in such rational functional form can be represented by Coxian distributions. Cox has shown that there is no increase in generality if additional feedback or feedforward arcs are introduced.

7. BCMP NETWORKS

Jackson networks are further generalized by introducing different classes of customers, and scheduling strategies other than FIFO at different service nodes of the network. The joint efforts of Baskett, Chandy, Muntz and Palacios have led into the product form solution of the network model referred with their initials (BCMP).

There are R classes of customers. The transition probabilities are denoted as $P_{ir,js}$ which expresses the probability that a customer in class r at service station i joins the queue at station j as a class s customer. Thus, it is possible to change classes as customers move through service stations. The job state is designated with a pair of indices (i, r) indicating its location and class respectively.

There are four types of nodes with different service disciplines in a BCMP network.

Type 1 node: There is a single server which serves according to exponential service times with the same rate $\mu_i(n_i)$ for all customer classes, and the rate is dependent on the number of customers n_i present at that station. The service discipline is FIFO.

Type 2 node: There is a single server with the server discipline as processor sharing. This means that if there are n customers requiring service, each is served at $1/n$ of the rate that a single customer is served by the service station. Thus, the service capacity is equally shared among the customers present without any consideration of the class type. The service time distribution is only restricted to have a rational Laplace transform. That is, they are required to be Coxian. They may have different parameters for different classes of customers.

Type 3 node: The service station acts as if there are infinite number of servers. This means that the number of servers is greater than or equal to the maximum number of customers that can be present at this service center. The service time distribution for each class of customers may differ, but they all have a rational Laplace transform.

Type 4 node: The service discipline is LIFO preemptive resume. That is, if the server is giving service to a customer when a new customer arrives, the service is interrupted and the new is served. The interrupted service is resumed after the new customer is finished being served. The service time distribution is Coxian as in the cases of 2 or 3 nodes.

Let the vector $s = (s_1, s_2, ..., s_K)$ denote the network state and $p(s)$ be the steady-state probability of being in state s. The global balance equations

$$p(s)[\text{Flow rate out of } s] = \sum_{s'} [\text{Flow rate out of } s' \text{ to } s] p(s')$$

with the normalizing equation

$$\sum_s p(s) = 1$$

may be employed to determine the steady-state distribution. Similar to the traffic equations in Jackson networks, the following set of equations

$$e_{js} = p_{0,js} + \sum_{i=1}^{K} \sum_{r=1}^{R} e_{ir} p_{ir,js}, \qquad i, j = 1, 2, ..., K \quad r, s = 1, 2, ..., R$$

must have a solution in order to obtain the steady-state distribution. The quantity e_{ir} is the relative frequency of the number of visits of class r customers into node i. The quantity $P_{0,js}$ is the probability of a new arrival joining node j as a class s customer.

BCMP theorem simply states that the general solution of the balance equations has the product form solution. That is, the joint probability is the product of the marginal probabilities which allows the network to be studied with each station taken in isolation.

The common property among the four node types described is that the Poisson input implies a Poisson departure process. This is referred as the "Markov implies Markov" property by Muntz (1972). These four node types are the only ones which are known to possess this property. Kelly (1976) introduced further generalization into the network models by allowing customers take arbitrary paths through the network rather than being governed by the transition probabilities $P_{ir,js}$.

8. REFERENCES

1. Barbour, A.D. 1976. Networks of Queues and the Method of Stages. *Adv. Appl. Prob.* **8**, 584-591.

2. Baskett, F., Chandy, K. M., Muntz, R. R. and Palacios, F. G. 1975. Open, Closed and Mixed Networks of Queues with Different Classes of Customers. *JACM* **22**, 248-260.

3. Burke, P. J 1956. The Output of a Queueing System. *Opns.Res.* **4**, 699-704.

4. Burke, P. J. 1976. Proof of a Conjecture on the Interarrival-Time Disribution in an $M/M/1$ Queue with Feedback. *IEEE Trans. on Comm.* **24**, 175-176.

5. Buzen, J. P. 1973. Computational Algorithms for Closed Queueing Networks with Exponential Servers. *CACM* **16**, 527-531.

6. Chandy, K. M. 1972. The Analysis and Solutions for General Queueing Networks. *Proceedings of the 6th Annual Princeton Conference on Information Science and Systems*. Princeton, New Jersey, 224-228.

7. Chandy, K. M., Howard, J. H. and Towsley, D. F. 1977. Product Form and Local Balance in Queueing Networks. *JACM* **24**, 250-263.

8. Cox, D. R. 1955. A Use of Complex Probabilities in the Theory of Stochastic Processes . *Proc. Cambridge Phil. Soc.* **51**, 313-319.

9. Çınlar, E. 1975. *Introduction to Stochastic Processes*. Prentice Hall, Englewood Cliffs, New Jersey.

10. Disney,R.L. and D. König 1985. Queueing Networks: A Survey of Their Random Processes. *SIAM Review* **27**, 335-403.

11. Gelenbe, E. and G. Pujolle 1987. *Introduction to Queueing Networks*. Wiley, New York.

12. Gordon, W. J. and Newell, G. F. 1967. Closed Queueing Systems with Exponential Servers. *Opns.Res.* **15**, 254-265.

13. Jackson, J. R. 1957. Networks of Waiting Lines. *Opns. Res.* **15**, 254-265.

14. Jackson, J. R. 1963. Jobshop-Like Queueing Systems. *Mgmt.Sci.* **10**, 131-142.

15. Jackson, R. R. P. 1954. Queueing Systems with Phase Type Service. *Opns.Res. Quart.* **5**, 109-120.

16. Kelly, F. P. 1976. Networks of Queues. *Adv. Appl. Prob.* **8**, 416-423.

17. Kiessler,P.C. and R.L. Disney 1988. Further Remarks on Queueing Network Theory. *Eur. J. Opnl.Res.* **36**, 285-296.

18. Kleinrock, L. 1976. *Queueing Systems: Computer Applications.* Vol 2. Wiley, New York.

19. Lemoine, A.J. 1977. Networks of Queues-A Survey of Equilibrium Analysis. *Mgmt Sci.* **24**, 464-481.

20. Melamed, B. 1982. Sojourn Times in Queueing Networks. *Math.Opns.Res.* **7**, 223-244.

21. Muntz, R. R. 1972. Poisson Departure Processes and Queueing Networks. IBM Research Report, RC 4145, IBM Thomas J. Watson Research Center, Yorktown Heights, New York.

22. Reich, E. 1957. Waiting Times When Queues are in Tandem. *Ann. Math. Stat.* **28**, 768-773.

23. Reiser, M. and Kobayashi, H. 1975. Queueing Networks with Multiple Closed Chains: Theory and Computational Algorithms. *IBM. J. Res. and Dev.* **19**, 283-294.

24. Sevcik, K. C. and Mitrani,I. 1979. The Disribution of Queueing Network States at Input and Output Instants.*Proceedings of the 4th International Symposium on Modeling and Perfecting Evaluations of Computer Systems.* North-Holland, Amsterdam.

25. Simon,B. and R.D. Foley 1979. Some Results on Sojourn Times in Acyclic Jackson Networks. *Mgmt Sci.* **25**, 1027-1034.

Chapter 6

QUEUEING THEORY
IN
COMPUTER SYSTEMS

Nihal Pekergin
Ecole des Hautes Etudes en Informatique
Université René Descartes
75006, Paris, France

1. COMPUTER SYSTEMS PERFORMANCE EVALUATION

All engineering systems are subject to performance evaluations: systems must be verified if they perform their intended functions and requirements. Correctness is seldom sufficient to make a system acceptable; a system must also be evaluated to see how well it performs its functions and requirements. Performance evaluation activities are always accompanied by some form of cost evaluations, a system having enough performance at a reasonable cost is considered satisfactory.

In computer systems, performance evaluation must be considered during the entire life cycle of the system (in design, development, configuration and tuning stages) and is of great importance. Computer system performance evaluation studies can be classified according to their objectives. A most popular classification divides these studies into three categories:

1. Design studies,
2. Improvement studies,
3. Selection studies.

In design studies, performance evaluation is undertaken by the producers of computer systems rather than its users. The goal is to reject terrible designs and to select a better one from performance viewpoint. Improvement studies are concerned with modifications made to an existing system in order to increase its performance or decrease its cost, or to improve both of them. These studies can also be called as optimization studies. A selection study tries to achieve certain objectives by choosing an existing system which is the most convenient for some specified selection criteria.

1.1. Performance Evaluation Methods of Computer Systems

Computer system performance evaluation methods are divided into two main areas: performance measurement and performance modeling methods (Ferrari (1978)). In performance modeling, resolutions are obtained by applying two classes of techniques: analytic techniques, and performance simulation techniques.

The obvious approach to system performance evaluation is to measure the performance directly, and this measurement can be realized with hardware dedicated and/or software dedicated tools. There are two major problems with performance measurement: first, measurement of most systems is a complex activity which involves considerable human and machine cost. Second, for systems which are not operational as in the design and development stages, measurement is not feasible. In such cases, performance measurement is intractable and the system performance can be only evaluated by performance modeling methods.

A performance model is an abstraction of the real system and it is developed to capture the main factors determining system performance. Once the model has been defined, it is parameterized to reflect any of the alternatives under study. Real system performance is then evaluated through the model.

Performance modeling provides quantitative predictions and also insights into the structure and behaviours of the real system. These are especially valuable in design stages. As a result of early recovery and elimination of terrible designs from performance viewpoint in these stages, an important amount of cost savings can be obtained. In design stages, performance models are simple and abstract; as more information becomes available, more detailed models can be developed. Once the system is measured, the previously developed models can be validated and modified.

Performance modeling is used not only in design and development stages. It is also applied in capacity planning and configuration. Capacity planning is the process of determining future computing system hardware needs based on projections of workload growth, where configuration is the reorganization of the material due to changes in workload. In these cases, performance modeling is inevitable since proposed systems are not operational.

1.2. Performance Modeling of Computer Systems

Computer systems can generally be characterized as a set of hardware and software resources, and a set of tasks competing for or using these resources. For hardware resources, we can cite processors, channels, memory partitions etc., while a lock for a database item, a compiler can be considered as software resources. Since there are multiple jobs for a limited number of resources, queueing for these resources are inevitable. Therefore, it is natural to model a computer system by a network of queues where customers are the jobs competing for the resources represented as service centers (stations). The characteristics of queueing networks such as resource utilizations, queue lengths and queueing delays are used to predict the performance of the real system through the model.

Research in performance modeling has been research in queueing theory. Indeed, queueing theory is first applied in communication systems, where it is still applied today.

Queueing network models are important tools in the design and analysis of computer systems, since they achieve a favorable balance between accuracy and efficiency for many applications. Queueing network models can be expected to be accurate to within 5 to 10 % for utilizations and throughput and to within 10 to 30 % for response times.

A class of models having mathematically tractable solutions, called product form queueing networks in queueing theory, have great importance in performance methodology. In analytic performance modeling, usually product form queueing networks are used as models (Gelenbe and Pujolle (1987)). Since these type of models are based on efficient solutions of mathematical equations, they are cost effective. So, they are widely applied and accepted performance evaluation techniques.

In order to obtain an analytic model, simplifying assumptions must be made, consequently analytic models cannot capture all details of the system. Nevertheless, many systems of interest may be analytically modeled with sufficient realism in order to provide insights into the parameters effecting system performance.

The studies on queueing theory may be summarized in three aspects.

1. Identification of classes of models which have mathematically tractable solutions,
2. Development of computationally efficient and numerically stable algorithms for product form networks,
3. Development of accurate and computationally efficient algorithms to approximate the solutions of large product form networks, and also non-product form networks.

When analytic performance modeling techniques have limitations on performance estimates, performance simulation modeling must be used. In simulation studies, systems may be modeled in any level of detail. The disadvantage of performance simulation modeling is that it is too long and expensive in terms of computing resources. However it is a very important tool in performance evaluation studies. Analytic models may be validated with simulation. Analytic models are also used to determine where a detailed analysis using performance simulation is required.

Today, performance simulation has been automated by simulation packages that generate the simulation program from a model description. However, they are expensive to parametrize, since detailed models require a large number of parameters. Running a simulation with a narrow confidence interval necessitates substantial computational resources. The cost of evaluating a queueing network by simulation often exceeds, by orders of magnitude, the evaluation cost of the same model by analysis. Nowadays, in performance evaluation studies, hybrid modeling methods which involve both simulation and analytic models, are largely applied whenever possible.

1.3. Parametrization of Performance Models

In computer performance modeling, parametrization of the model is very important. It is evident that a performance evaluation study can be no more accurate than the parameter values provided to the model.

The input parameters of a queueing network model can be divided into three groups:

1. Customer description,
2. Service center description,
3. Service requirements.

Customer description is the workload, characterization of the underlying system. Three workload types are considered: batch workload, interactive workload and transaction workload. For interactive and batch workloads, closed queueing network models are used, so the number of jobs in the system must be stated in order to define the workload intensity. Additionally, for an interactive workload, interaction time spent by thinking and/or typing between interactions must also be defined. Computer system with transaction workloads are modeled by open queueing networks, in these cases the arrival process of transactions to the system must be introduced to the model as a parameter to define the workload intensity.

Customers are partitioned into classes. Within a class, all customers are homogeneous, but different customer classes may have different properties. It is possible to model computer systems with different workload components by defining multiple customer classes.

To define a customer class, the transition properties (routing) between service centers and the service requirements in each center must be specified. Service requirements are represented by random variables. For an existing system, measurements and statistical data analysis techniques are used to determine the distribution functions, and synthetic workloads are used to parametrize proposed models.

Center description determines the correspondence between the system resources and the service centers in the queueing model. Two types of service centers are defined: queueing centers, and delay centers. Delay servers are the service centers having only service time delays but not queueing delays, and they can be considered as queueing centers with infinite servers.

The parameters that must be described for a service center are the number of servers which are generally assumed to be identical, the storage capacity or the maximum number of customers that may be admitted to the station which is generally taken to be infinite. Another important description is that of queueing disciplines, this parameter describes the order in which the customers are taken from the queue and allowed into service. Standard queueing disciplines are First-Come-First-Served (FIFO), Last-Come-First-Served (LIFO) with preemption or not, Processor Sharing (PS) and priority disciplines in which priority among classes may be established.

1.4. Computer System Performance Measures

Since performance is a subjective concept, its definition may differ for different categories of people (manufacturers, system programmers, application programmers, etc.). There are also the performance measures which are difficult or impossible to quantify. For example, the ease of use of a system is a performance measure which is difficult to quantify. We are interested in quantitative performance measures (Lazowska et al.(1984)). The standard performance measures can be classified according to their properties as follows:

1. Productivity: This is defined as the volume of information processed by the system in unit time. The performance measures used to evaluate the productivity may be system throughput (the number of customers served in the system during the unit time) or system capacity (maximum system throughput). Evidently, the productivity of the system must be high to have better performance.

2. Responsiveness: This is defined as the time between the presentation of an input to the system and the appearance of the corresponding output. The response time of the whole system or any of its subsystems may be performance measures of interest. From performance viewpoint, the responsiveness of the system must be fast, especially for real-time systems.

3. Utilization: It is the ratio of time in which a specified part of the system is used (or used for some specified purposes) during a given time interval. Resource utilizations are important performance measures. In queueing theory, we are generally interested in the steady-state (limiting) behaviour of the system. This measure is especially useful to determine the bottlenecks of the system, which are the resource that have the highest utilizations.

Little's theorem is the principal theorem of queueing theory which statres the relationship between performance measures in steady-state. If L is the average number of customers, W is the response time, and λ is the throughput, then

$$L = \lambda W \ .$$

This is a general result which does not depend upon any specific assumptions regarding service time, arrival distributions, the number of servers, and the queueing disciplines. Moreover, this theorem is applicable in any level of the system; that means, for the whole system, or for any subsystem, station, or server.

2. COMPUTER SYSTEM MODELS

2.1. Central Server Model

A simple queueing network model of batch systems called central server model is given in Figure 1. This model is also referred to as cyclic queue model because of the cycling of the programs between two queues, one representing the CPU while the other one representing one of the I/O devices.

A batch program is considered to consist of a number of CPU-I/O cycles. It executes on CPU, performs an I/O operation on one of the I/O devices, then returns to the CPU, and so on. This process is repeated until the program is completed. This model is first proposed to estimate the system throughput and device utilizations of an IBM 360/75 (Trivedi (1982)).

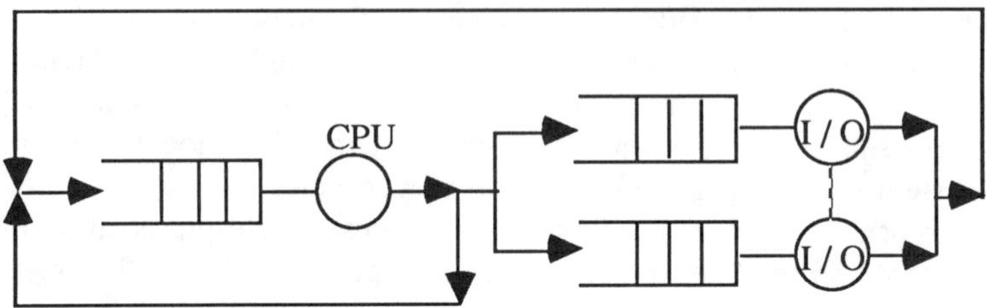

Figure 1. Central Server Model for Batch Systems

In order to define the model completely, the following must be specified:

1. The number of jobs (batch programs) in the system.
2. The scheduling disciplines of jobs at each center.
3. The service requirements of jobs at the centers.

A number of implicit assumptions on the structure of the model are made as follows:

1. The multiprogramming degree is constant; whenever a program finishes, it is replaced immediately by another one.
2. The programs behave homogeneously, we cannot distinguish their behaviours.
3. When a job leaves CPU, it proceeds to the I/O device i with the probability p_i.
4. The service times at each center are independent random variables.

The solution of this model depends on its characterization as a Markov process. The parameters of the model are chosen so that a product form solution may be possible. Hence, scheduling disciplines and service time distributions at each center are assumed so that the BCMP theorem can be applied.

In the original model, the service times are chosen as exponentially distributed random variables, and the scheduling disciplines is First-Come-First-Served. Therefore, four parameters: the number of jobs, the average CPU service time, the number of disks, and the average disk service time are used to describe this model. The solution of the model can be found with computation algorithms for closed queueing networks (Lavenberg (1983), Sauer and Chandy (1981)).

2.2. A Time-Sharing System Model

In a time-sharing system, a job is called active when it is admitted into the main memory. The degree of multiprogramming is defined as the number of active jobs at a given time. If the degree of multiprogramming is too low, then the system is underutilized. On the other hand, if it is too high, then a small portion of memory is allocated to each active job. Therefore, the pagination actives increase so that a little useful work is done. This phenomena is called thrashing . The multiprogramming degree is limited in a time-sharing system, the problem is then to find a multiprogramming degree which provides a higher utilization without thrashing.

Main memory is allocated to the active jobs in units of fixed size called page frames. Every job requires a certain set of frames in order to complete its execution. This set is referred to as its virtual memory. Only a fraction of a job's virtual memory can be contained in the page frames allocated to that job, the rest is stored in secondary memory. When an active job requires a page from its virtual memory which is not already in main memory, a page fault occurs. Then, a new page frame has to be allocated to this job. In turn, this may necessitate the freeing of another page from the memory allocation for that or another job.

Main memory partition influences system performance through the page faults, the traffic of page to and from the paging devices, and the ensuing delays. These effects can be incorporated into the model in a convenient way by associating with each active job a lifetime function , $e(m)$. For a job that is executing in m page frames of main memory, $e(m)$ is the average amount of CPU service received by the job between two consecutive page faults. Thus, $1/e(m)$ can be defined as the rate at which a job having m page frames interrupts its CPU service to go to the paging device.

Intuitively $e(m)$ should be an increasing function of m; if a job has more page frames, then page faults occurs less frequently. Generally, two analytic forms of $e(m)$ have been used (Gelenbe and Mitrani (1980)):

1. Belady-Kuehner Formula:

$$e(m) \; = \; am^k \qquad a > 0, \; k > 1 \; .$$

2. Chamberlin-Fuller-Lin Formula:

$$e(m) \; = \; \frac{2b}{1 + (c/m)^2} \qquad b, c > 0 \; .$$

In Figure 2, a queueing network model for a time-sharing system with a limited multiprogramming degree is given. Thinking and/or typing states of interactive users (terminals) are modeled by delay servers. In such systems, there is simultaneous resource possession; a job must gain admission to the multiprogramming set before competing for active resources such as processors. A memory partition can be considered as a passive resource. The jobs which can not have a memory partition must wait in the memory queue.

Modeling simultaneous or overlapped possession of resources in queueing networks poses difficulties since certain essential assumptions to have a product form solution are violated. Therefore, efficient exact solution methods are not applicable, and approximate solutions must be applied. The approximate performance estimates of this model can be found in a hierarchical manner, by decomposing the model into two components as given in Figure 3.

The interactive users represented by delay servers will be called the outer system. The subsystem which contains the processor, I/O devices and the memory queue will be called the inner system. The inner system is represented by a flow-equivalent server. The decomposition is feasible due to the fact that interactions between the inner and the outer system are weak compared to those within the inner system.

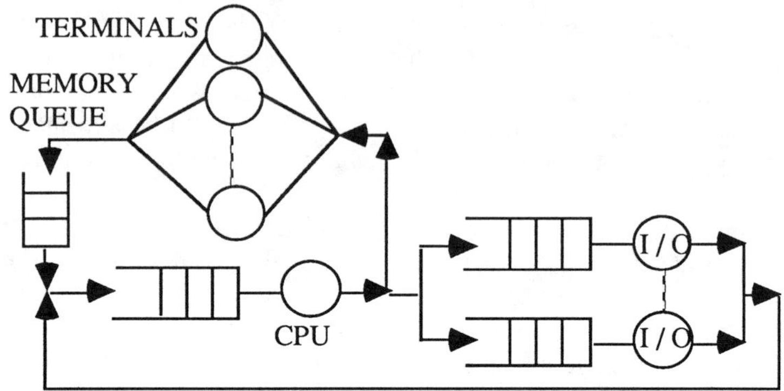

Figure 2. A Memory Contention Model for a Time-Sharing System

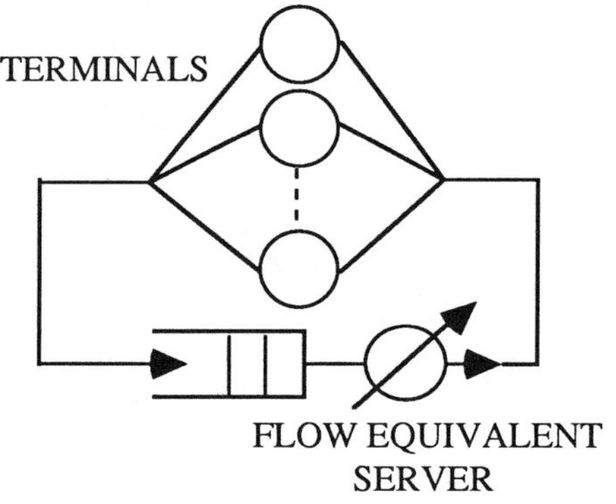

Figure 3. High Level Model of the Memory Contention Time-Sharing System

It is assumed that the inner system reaches equilibrium between consecutive changes in multiprogramming degree, so the system may be modeled by a closed queueing network given in Figure 4.

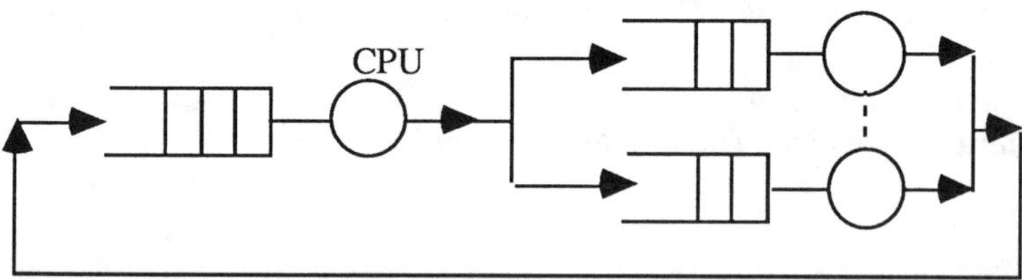

Figure 4. Model for the Inner System

First, the steady-state throughput of the model for the inner system is calculated for each possible customer population (maximum multiprogramming degree). Then, the load dependent service rate of the flow equivalent server in the higher level model in Figure 3 is considered to be equal to the throughput of the inner system in this load. Therefore, the system response time is calculated through the higher level model. The multiprogramming degree which allows high productivity and sufficiently short response time may be determined from this analysis.

An algorithm to evaluate system response time of such a system is given in Table 1. In this algorithm, Mean Value Analysis is used and the paging fault rate is represented by the Belady formula.

Table 1. Approximate Solution of a Time-Sharing System (Pagination Characterized by Belady Formula)

```
const  N        :Active terminal number;
       K        :Maximum multiprogramming degree;
       M        :Memory partition number;

var    i        :Multiprogramming degree;
       R(i)     :Mean inner subsystem response time;
       X(i)     :Inner subsystem throughput;
       V        :Total I/O devices visit number;
       V_CPU    :CPU visit number;
       D_CPU    :Total CPU time per customer;
       T        :Mean service rate;
       a, k     :Belady parameters;
begin
for i := 1 to K do
   begin
   V_CPU := V + 1 + D_CPU / a · (M/i)^k;
   MVA-SOLVE(BATCH-SYSTEM(CPU,DISKS) with POPULATION = i);
   X(i) := i / R(i);
   end;

for i := K + 1 to N do X(i) := X(K);
    MVA-SOLVE(HIGH-LEVEL-MODEL (TERMINALS, EQUIVALENT-SERVER)
    with T_EQUIVALENT-SERVER(i) := X(i));
end;
```

2.3. Simultaneous Resource Possession

In computer systems, simultaneous resource possession may arise in different contexts: in systems with limited multiprogramming degree, in I/O subsystems, in multi-processor system, etc. As mentioned above, simultaneous or overlapped possession of resources in queueing networks violates essential assumptions to have a product form solution.To evaluate the performance of such systems method of surrogates can be used (Jacobson and Lazowska (1982)).

The main idea in this method is to partition queueing delay according to simultaneously held resources which are responsible for the delay. This is an important point to determine and eliminate the bottleneck of the system. In such systems, primary resources are defined as the resources held for a preliminary service time, possibly zero, and while using the resources of the secondary subsystem.

In systems with limited multiprogramming level, a memory partition constitutes the primary resource, and the CPU and I/O subsystem constitute the secondary subsystem.In an I/O subsystem, a disk constitutes the primary resource where the channel is the secondary subsystem.In a multiprocessor system which contains a number of buses to connect the memory modules to the processors, a bus is a primary resource while memory modules constitute the secondary subsystem.

In this method, two models are constructed: one of them includes primary resources explicitly, and a delay server acting as a surrogate for queueing delay due to congestion in secondary resources. The second model consists of an explicit representation of secondary resources, and a delay server to surrogate the queueing delay due to congestion in primary resources. The solutions of these two models are iterated in order to partition the queueing delay between simultaneous resources.

Since primary resources pose a population constraint in the secondary subsystem, a flow-equivalent representation must be used for this subsystem, an analysis similar to that of Section 2.2. can be applied. To analyze the secondary subsystem in isolation, an augmented secondary subsystem model which consists of secondary subsystem resources and a delay server whose service time is equal to the nonoverlapped primary resource service requirement is established, and the thoughput of this model is calculated for all numbers of customers (number of primary resources). The service rate of the secondary subsystem for a given load is considered to be equal to the throughput of the augmented subsystem in this load.

We now present, an example of the application of the method of surrogates. As mentioned above , this method may be applied to determine the disk and the channel contention in a computer system. The system is considered to be composed of a CPU, a channel, and a number of disks which are non-RPS (Rotational Position Sensing),which means they are capable of independent seeking but not capable of rotational position sensing . Then, the channel is occupied when a disk is either searching or transferring the data. In this case, the disks are the primary resources and the channel constitutes the secondary subsystem.

The model which can be called the disk contention model contains the CPU, the disks, and a delay server representing the channel queueing delay as given in Figure 5. The channel contention model in Figure 6 contains the CPU, and the flow-equivalent server representation of the channel. In Figure 7, the augmented secondary subsystem model which is used to determine the flow-equivalent server representation of the channel is given. In this subsystem, nonoverlapped primary service requirement is the disk occupation for seeking.

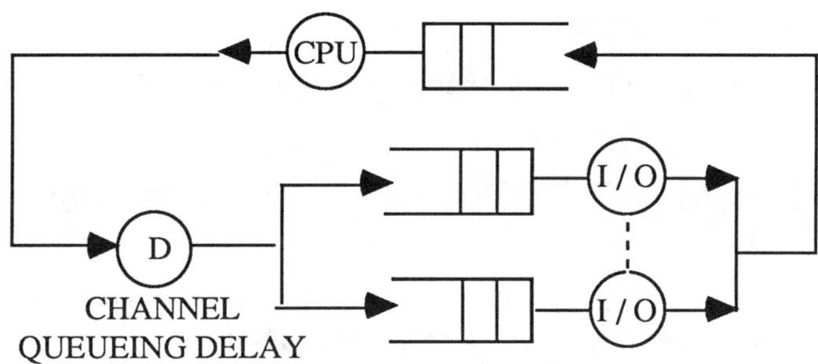

Figure 5. Disk Contention Model

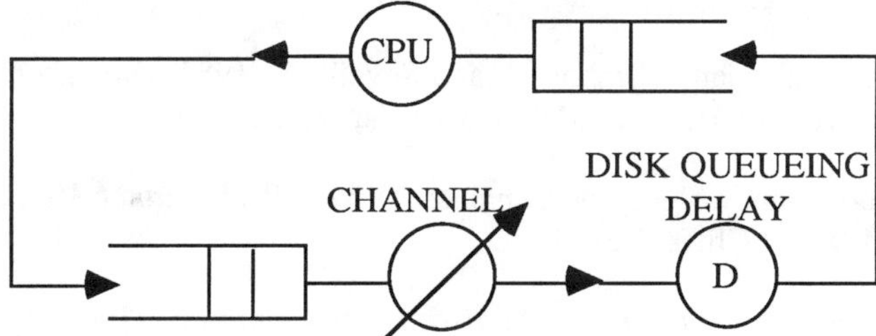

Figure 6. Channel Contention Model

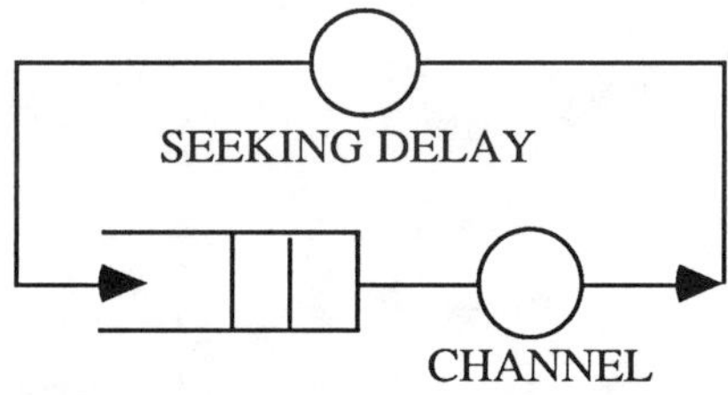

Figure 7. Augmented Secondary Subsystem Model

The disk contention model is solved with a zero channel queueing delay, and the disk queueing delay calculated in this model is used in the channel contention model. This process is iterated until queueing delays do not change significantly in both of the models. These limiting results give the approximate queueing delays for disks and the channel.

3. REFERENCES

1. Ferrari,D. 1978. *Computer System Performance Evaluation.* Prentice-Hall, Englewood Cliffs, New Jersey.

2. Gelenbe,E.,and Mitrani,I. 1980. *Analysis and Synthesis of Computer Systems.* Academic Press, New York.

3. Gelenbe,E.,and Pujolle,G. 1987. *Introduction to Queueing Networks.* John Wiley, New York.

4. Jacobson,P.A.,and Lazowska,E.D. 1982. Analyzing Queueing Networks with Simultaneous Resource Possession. *CACM* **25**, 142-151.

5. Lavenberg,S.(Ed.) 1983. *Computer Performance Modeling Handbook.* Academic Press, New York.

6. Lazowska,D.,Zahorjan,J.,Graham,S.,and Sevcik,K. 1984. *Quantitative System Performance.* Prentice-Hall, Englewood Cliffs, New Jersey.

7. Sauer,C.,and Chandy,M. 1981. *Computer System Performance Modeling.* Prentice-Hall, Englewood Cliffs, New Jersey.

8. Trivedi,S. 1982. *Probability and Statistics with Reliability, Queueing and Computer Science Applications.* Prentice-Hall, Englewood Cliffs, New Jersey.

Chapter 7

SIMULATION OF QUEUEING MODELS
IN
COMPUTER SYSTEMS

M.S. Obaidat
Department of Electrical and Computer Engineering
University of Missouri-Columbia/Kansas City
Independence, MO 64050, USA

1. INTRODUCTION AND RATIONALE

There are three main techniques used to evaluate the performance of computer systems. These are performance measurement, analytic queueing performance modeling, and simulation performance modeling (Heidelberger and Lavenberg (1984)). If a prototype or an existing computer system is available, it may be tested under various load conditions (i.e., different traffic load conditions, CPU- bound and R/W-bound, and moderate jobs, etc.), and by studying the measurements, the performance of the computer system can be known and suggestions for future expansion and planning can be made. This technique is considered the most accurate method, but it is the most expensive and often not feasible in the capacity planning and design stages when usually no real hardware is available (Sauer and Mac Navir (1983)).

The second method is the analytic queueing approach which is the simplest, fastest and least cost approach. Once the queueing solution is available, different parameter values may be substituted into the expression to find the required performance measure with little effort. Unfortunately, queueing modeling is usually inaccurate due to the various assumptions considered in the analysis to find closed-form solutions (Sauer and Mac Navir (1983)).

The third technique, the simulation technique, is used when accurate and transient results are desired for a computer system that does not exist, or even when specific simulation experiments are required to be performed on an already existing system (Mac-Dougall (1987)). In such cases, event driven computer simulation is used as a mean for performance evaluation of computer and data communication systems.

Simulation in general, means using computers to imitate the activities in a system, like a computer system, so as to study the behaviour and performance of that system. The accuracy of the simulation model depends, among other factors, on how close the model imitates the real system. There are several types of simulations that have considerable real-world importance. These include continuous, discrete-event, and Monte Carlo simulations.

Discrete-event simulation has been used heavily to model computer systems in recent days since it is concerned with modeling of the system as it evolves over time by representation in which state variables change only at a specific number of points in time (Law and Kelton (1982)). Therefore, we can characterize discrete-event simulation by:

1. State changes occur at specific finite number of times,
2. State changes are instantaneous.

In general, a simulation model is one laboratory version of a system. Once the model is developed, experiments can be conducted on. Such experiments or simulation runs enable us to derive some conclusions about the future of the project to be built or the modification to be performed to enhance the already existing system. These experiments can be conducted:

1. Without constructing the suggested computer system,
2. Without disturbing the already existing system if it is operational, which is costly and sometimes unsafe to experiment with,
3. Without damaging the computer system if the goal is to determine the worst case of operation.

Obviously, simulation modeling of computer systems can be used for the design, analysis and performance prediction and assessment.

Continuous-time simulation is used to model computer systems at the circuit level in which state switching behaviour is analyzed. At the gate-level simulation, the components, transistors, resistor etc. of the circuit are aggregated into a single element. Sets of gates are aggregated into elements such as multiplexers, adders in the register-transfer level. System-level simulation starts at a level higher than the register-transfer level. The system-level modeling is developed in order to analyse the system from the viewpoint of performance and usually it represents only those elements related to the performance evaluation issue.

The dynamic composition of a computer system can be described in terms of activities,processes and events. In discrete simulation, the state of the system can change only at event times, and a complete dynamic portrayal of the state of the system can be obtained by advancing simulation time from one event to the next. A discrete simulation model may be developed by defining the changes in state that happens at each event time by :

1. Specifying the activities in which the entities in the system engage,
2. Describing the process through which the entities flow.

Figure 1 shows the relationship between events, activities and process . It is common to call these three approaches as events, activity scanning and process oriented discrete simulation techniques.

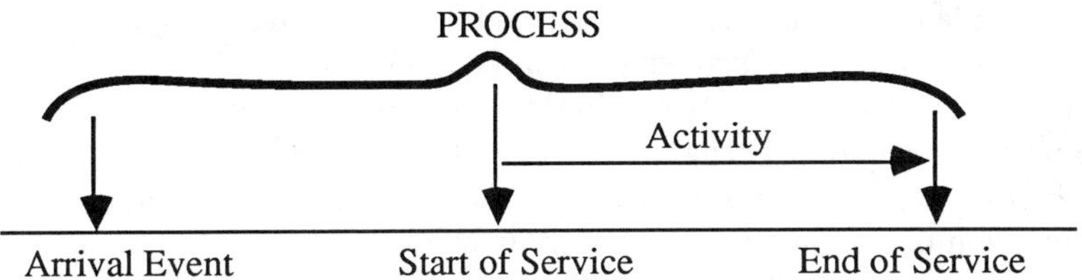

Figure 1. Relationship Between Event, Activity, and Process

2. MODEL REPRESENTATION

In order to represent queueing models of computer systems two tasks are required: system representation and work representation.The two tasks are interrelated since the level of abstraction at which work load is represented must correspond to that at which the system is represented.The task of describing the work done by a specific system is called "work load characteristic" (Fishman (1978),Merle et al. (1987)). The majority of the case studies considered in this chapter center around contention for the use of computer system resources (servers).This contention phenomenon causes work to be queued for execution. This may affect the performance of the computer system. The distributions of the appropriate variables and their mean values must be considered when the work load of the system is described.If the system to be modeled by simulation is existing then we may be able to measure and obtain real distributions by the use of some electronic devices such as hardware and software monitors, counters, logic analysers, etc. After the input parameters are collected, we can do a table look-up to obtain a samples value or we can fit a theoretical distribution, such as exponential or Gamma etc., to the actual distribution collected by the aforementioned methods. An algorithm can be devised to generate a sample value. There are off-the shelf computer programs that enable us to substantially reduce the time and effort required to find a best fit distribution. The program developed by Law and Vincet (1983) allows a variety of data transformations.This program, called UNIFIT, fits up to nine different distributions simultaneously and it offers graphics displays of the results.

In order to model a new architecture, and in such cases the system is not built yet, it is possible to adapt a distribution that is approximated by conducting empirical study on some existing system or a portable one. For instance, if we want to model a new computer system and it is required to specify the distribution of the CPU compute (execute) intervals; then we can randomize the number of instructions on the current system executed per interval to be a representation of the real system.

One good example used by the author to represent the work load of the Image MultiProcessor System (IMPS) is described below. The processor used in the architecture is MC68000 16/32 bit microprocessor. Since we had one uniprocessor module based on this processor, we ran a spectrum of the application programs on this uniprocessor. The status signals (function code outputs) FC0, FC1 and FC2 were connected to the logic analyser probes along with the address lines, chip select, R/W, AS and DTAK signals.

The function code outputs indicate the state and the cycle type currently being executed. The data sheets of the manufacturer (Motorola(1983)), shows that when the values of FC2, FC1, FC0 are (010) or (110) in binary and AS is asserted, then the processor is using the program memory. Further, if the R/W signal is high then the operation is read. Thus, if the two previous conditions are satisfied (i.e., program memory is being used and the operation is read) then the processor is executing code for a duration of time we called the execution (compute) time. These values of computer times are used as inputs to the IMPS simulation (Obaidat (1989)).

3. DIFFICULTIES OF COMPUTER QUEUEING MODELS' SIMULATION

There are two main problems with computer queueing models' simulation: level of details in the model and computation time consumption. It is important to be able to adjust the length of simulation run in order to obtain estimates of accepted accuracy. The problem of accuracy assesment and simulation length can be studied by the use of confidence intervals. Assume θ is some unknown characteristic of the model to be estimated, such as the mean response time. An interval (K, L) with random end points K and L is said to be $100(1-\alpha)\%$ confidence interval for θ if $P\{K \leq \theta \leq L\} = 1 - \alpha$.

For $(1 - \alpha)$ close to one, there is a high probability that the parameter θ is between K and L. The confidence interval is usually formed in a way that a point estimate $\hat{\theta}$ of θ is frequently around the midpoint of the interval. The width of the interval (K, L) is a measure of the accuracy of $\hat{\theta}$; the narrower this interval is, the more confidence we have in the estimate.

The other main problem is the difficulty to create the simulation program and verify its correctness. If general-purpose high-level languages are used to code the simulation program, then the programmer is responsible for the coding supporting routines such as the random number generator, simulation result collector, etc. in addition to the procedures required to represent the model itself.

The development of simulation models is greatly benefited from the use of simulation languages such as SLAM, GASP, GPSS and SIMSCRIPT ll.5 (Russell (1983), Schriber (1974), O'Donovan (1979)). These simulation languages automatically handle the supporting procedures so that programmers can spend more time on the model itself. Moreover, they provide high level constructs and features common to all simulations such as event scheduling, statistics reports and gathering, queue management, and of course, random number generation. Most of these simulation languages have a pc-version in addition to versions that run on almost all mainframe and supermini computers. Simulation packages that are specifically designed for computer performance evaluation have also been introduced. These computer performance evaluation oriented packages provide high level modeling structures that are well suited for simulation of queueing network models of computer performance. Examples of such packages include QNAP, RESQ, PAWS, PANACEA, MAP, and NETWORK ll.5 (Heidelberger and Lavenberg (1984), MacDougall (1987), Fishman (1978), Merle, et al. (1987)). Furthemore, high level packages exist for capacity planning such as SNAP/SHOT (MacDougall (1987)) that includes built in models of certain IBM devices and control programs. These simulation packages minimize the time required for learning the simulation languages since a considerable amount of time is needed to master and become efficient in using them. Moreover, system designers often do not have the time to acquire the knowledge in these simulation languages. Therefore, simulation packages such as the aforementioned ones simplify the programming process and reduces the time required to build and debug the simulation programs. However, these packages are not flexible and may have some redundancy.

4. SIMULATION LANGUAGES FOR COMPUTER SYSTEMS

The prevalent use of simulation as an analysis and prediction tool has led to the development of a number of languages that are simulation oriented languages. These simulation languages provide specific statements and options for representing the system at one point in time and moving the state of the system from one state to another. It is not the purpose of this section to provide a manual to all simulation languages, but to illustrate some differences and similarities between most commonly used languages.

The main advantages of simulation languages over general purpose languages can be enumerated as below:

1. Simulation languages decrease the amount of programming time since they provide features that are utilised in the development of models.
2. They have basic building blocks that are akin to simulation than general purpose languages.
3. Simulation languages provide better error detection.
4. Models written in simulation languages are more flexible than those written in general purpose languages.
5. Some simulation languages such as SIMSCRIPT ll.5 have the advantage of providing dynamic storage allocation during execution (Russell (1983), CACI (1983), Gordon (1975)).

Simulation languages could be classified into process oriented, and continuous simulation languages. Discussion on GPSS and SIMSCRIPT ll.5 will be presented in this section as examples on two different simulation languages. Moreover, NETWORK ll.5 and COMNET II.5 will be discussed as an example of the simulation packages developed for computer queueing models simulation.

4.1. GPSS

GPSS (General Purpose Simulation System) is a process oriented simulation language. One well known version of the GPSS is the GPSS/360 version (Schriber (1974), O'Donovan (1979), Gordon (1975)). GPSS is appealing due to its simplicity. The model developed by the use of GPSS is constructed by combining standart blocks into a block diagram that represent the logical structure of the system. In order to code GPSS programs, we need to learn the functional steps that logically combine these blocks to represent the Computer System of concern. GPSS is limited in computing power and does not have floating point and real arithmetic capabilities. This means that GPSS simulation clock is of integer value, i.e., time between arrivals and service time for customers must be an integer number of units. There are more than forty building blocks in GPSS. Figure 2 shows some of the basic building blocks of GPSS.

In order to execute the program by the GPSS processor, the block diagram must be translated into an equivalent statement for execution. Every block has three types of information: a symbolic location name, the operation, and the block's operands. Transactions in GPSS are created by the GENERATE block with the operands A, B, C, D, E, F and G. The A operand specifies the mean time between creations. If B is specified as a number, then the time between creations is distributed uniformly in the range from (A-B) to (A+B). The C operand specifies the time of the first transaction and is referred to as the offset interval. The D operand determines a limit on the number of transactions that can enter the model by the GENERATE block. The priority of the transaction is determined by the E operand that has associated parameters. The TERMINATE block is used to destroy transactions. Moreover, the TERMINATE block is also used to reduce the termination counter by the value specified as its A operand (Schriber (1974), O'Donovan (1979), Gordon (1975)).

The ADVANCE block is used to provide the time advance. When a specific transaction enters the ADVANCE block, its progress is delayed by the time specified by the operands A and B. The definitions of A and B operands are made by the same way the A and B operands of the GENERATE block. The A operand prescribes the average delay time and the B operand defines the half-width of a uniform distribution.

A user-written table function must be included in the model in order to generate samples from distributions other than the uniform distribution. The required function is defined by a function header card and at least one more card. This function header describes the name of the function, the number of points specified for the distribution, and the random number stream. The other cards that follow this function card contain the entries $/X_1, Y_1/X_2, Y_2/.../X_n, Y_n/$ where X_i and Y_i are the i'th cumulative probability and associated function value, respectively. Resources in GPSS are modeled as either storages or facilities. Facilities may be preempted using PREEMPT block whereas storages cannot. Moreover, facilities can only be released by the transaction which seized it. The ENTER and LEAVE blocks are used for capturing and freeing storages. The QUEUE and DEPART blocks are used to obtain statistics on waiting and queue length. The GPSS processor maintains certain variables that describe the system status during simulation. The ASSIGN is used to block assign values to parameters of the transactions. Figure 3 shows a GPSS model of a single-server queueing system and Figure 4 shows the GPSS coding of this model (Schriber (1974), O'Donovan (1979), Gordon (1975)).

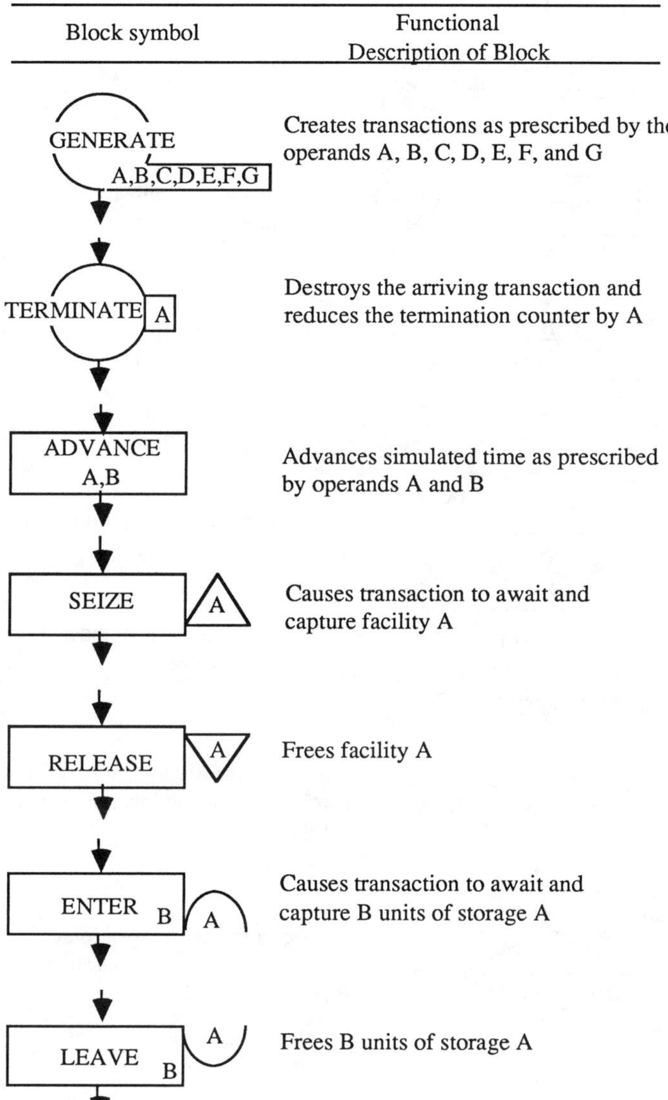

Figure 2. Basic Building Blocks of GPSS Simulation Language

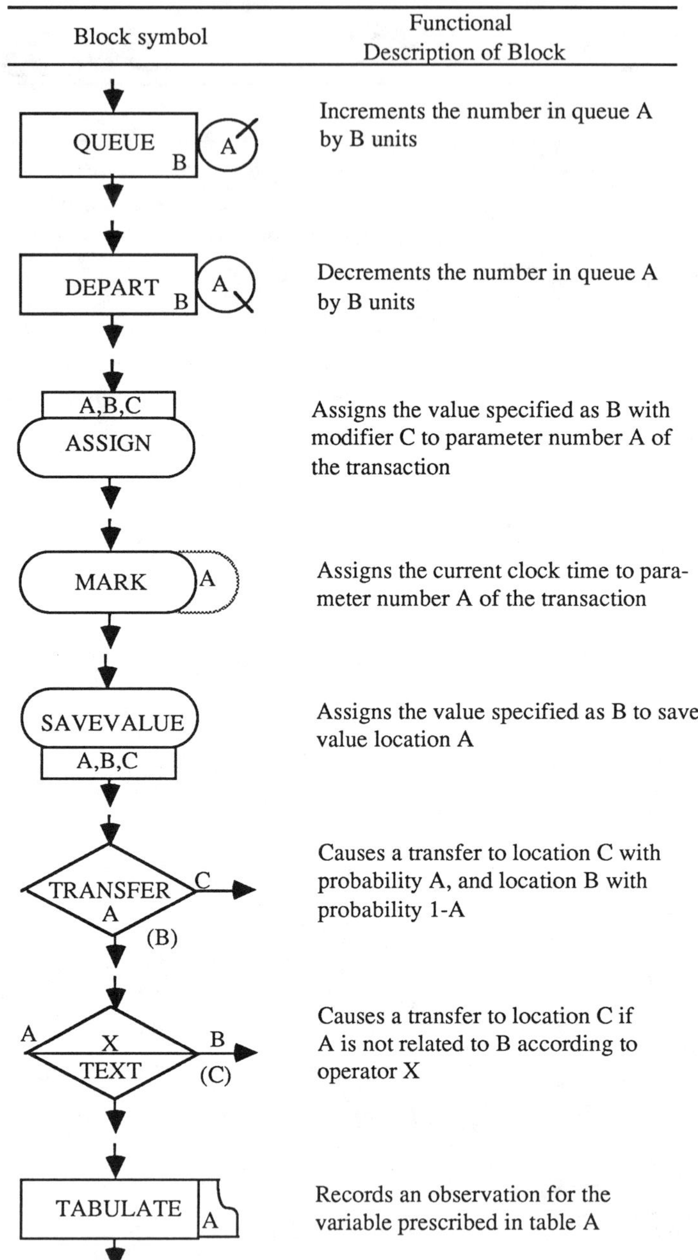

Figure 2. (Continued) Basic Building Blocks of GPSS Simulation Language

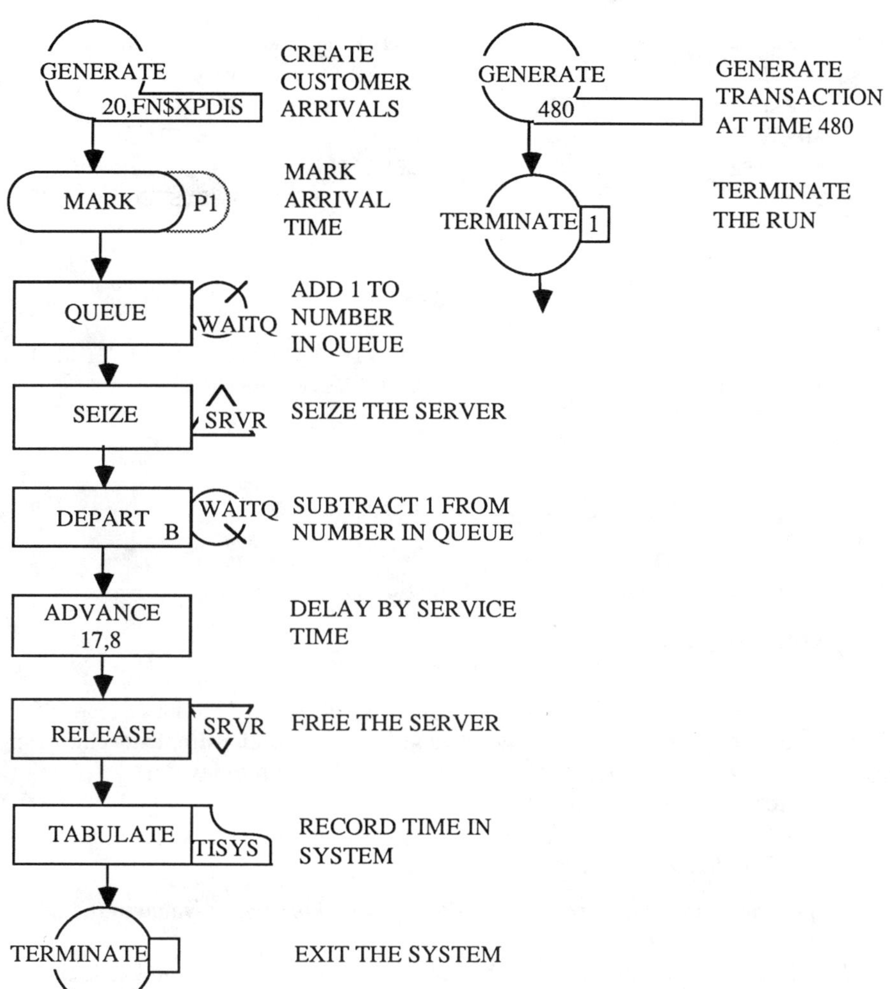

Figure 3. A GPSS Model of a Single Server Queueing System

```
TISYS TABLE         MP1,0,5,20

        MODEL SEGMENT

        GENERATE        20,FN$XPDIS      CREATE CUSTOMER ARRIVALS
        MARK            P1               MARK ARRIVAL TIME
        QUEUE           WAITQ            ENTER THE WAITING LINE
        SEIZE           SRVR             SEIZE THE SERVER
        DEPART          WAITQ            EXIT THE WAITING LINE
        ADVANCE         17,8             DELAY BY SERVICE TIME
        RELEASE         SRVR             FREE THE SERVER
        TABULATE        TISYS            RECORD TIME IN SYSTEM
        TERMINATE                        EXIT


        TIMING SEGMENT

        GENERATE        480              CREATE TRANSACTION AT TIME 480
        TERMINATE       1                TERMINATE THE RUN


        CONTROL CARDS

        START           1                START THE RUN
        END
```

Figure 4. GPSS Program Listing of the Single Server System

4.2. SIMSCRIPT ll.5

SIMSCRIPT ll.5 simulation language was developed by Kiviat et al. (1969). The history of Simscript ll.5 consists of five stages. It started from a simple teaching language designed to introduce programming concepts at level one up to level five. Russell (1983) gives a history of this simulation language.

SIMSCRIPT ll.5 has the following main advantages over some other simulation languages (CACI (1983), Kiviat et al. (1969)):

1. Its programs are easy to read and to debug, which is of great value in the developing stages.
2. It is clear and naturally reflects the logic of the system to be modeled, since it provides powerful capabilities to describe the system under consideration for modeling.
3. SIMSCRIPT ll.5 models are top down oriented through a PREAMBLE in which system elements and their relationships, statistics to be obtained are all defined before writing the executable code.
4. Programs written in SIMSCRIPT ll.5 are easy to be modified by the programmers who did not write the first version since it is English like!

Every SIMSCRIPT ll.5 program begins with a PREAMBLE whose statements are of declarative nature. It is used to define all the building blocks for the simulation model. Moreover, it is used to define the global variables, unit of time for the simulation clock, and the required measures of the system performance such as average delay in queue, number of jobs done, throughput, utilization of resources, etc. The main program is the core of the SIMSCRIPT ll.5 simulation program since it is where the execution begins, and all routines are called. The general structure of the MAIN program is as shown below:

```
MAIN
   CALL READ.DATA
   CALL INITIALIZE
      :
   START SIMULATION
END
```

As shown above, the execution of the MAIN program begins by calling READ. DATA which is not a process routine. It is used to read and print input parameters for the simulation. The INITIALIZE routine is called to initialize specific variables and place the "Initial" process notices, into the event list. The simulation operation begins with the execution of the START SIMULATION statement since this calls the system routine. When the event list contains no more process notices, then the END statement is executed, and this results in the termination of the simulation. Moreover, a STOP statement terminates whenever it is executed in some routine. There will be a process routine for every process defined in the PREAMBLE. The process and its corresponding routine must have exactly the same name. A process notice is a representation of a process entity that is present in the simulation model. This process notice contains an activation time with an associated activation point. In order to compute and print the simulation results, each simulation model should have a REPORT routine. The location of the call statement to this routine and whether it is considered as a process or a regular routine depends on the stopping rule for the simulation.

There are two important buildings in SIMSCRIPT ll.5: processes and resources (Russell (1983)).

1. Processes

All processes are defined in the PREAMBLE and they should have a corresponding process routine. The realization of a process entity comes into existence when the ACTIVE statement is executed. One of the system attributes is TIME.A which is the activation time for the notice. Each type of process defined in the PREAMBLE has an associated global pointer that has the same name. A process notice could be at any instant of time in one of three possible states: pending, delayed for resource or created. The RQUEST statement is used by a specific process entity in its process routine to request a certain number of units of a resource. If the required number of units are available, then they are made busy, and process entity moves to the next statement in the process routine. However, if the units are not available then the entity is put in the queue and control is given to the timing routine.

The RELINQUISH statement is used to relinquish a certain number of units of a resource of a specified type. The WORK/WAIT statements are used by some process entity to place its process notices back into the event list for an amount of time. The WORK statement is generally used to represent service times while the WAIT statement is used to represent interarrival times. Control is returned to the timing routine after WORK/WAIT statements. The SUSPEND statement is used by a specific process entity to stop its execution of the corresponding process routine. The notice will stay until reactivated by a RESUME or REACTIVATE statements. The ACTIVATE and REAC-TIVATE are actually exchangeable, however the ACTIVATE is used to create a new process notice.

The INTERRUPT statement is used to remove a process notice from the event list even if the interrupted notice is in the middle of a WORK or WAIT statement. The RESUME statement is used mainly in a process or normal routine to resume an interrupted process notice. The CREATE statement is used to create a process notice for a specific type of process without placing it into the event list. In order to destroy a process notice the DESTROY statement is used. At the end of each process routine the END statement is used. The RETURN statement, if placed in the middle of a process routine, has the same effect as the END statement.

2. Resources

The service provided by an object such as a processor or a machine is called a resource. The resource is defined in the PREAMBLE. Allocation of storage for a resource is made by the CREATE statement. For more complex SIMSCRPIT ll.5 data structure, the reader is referred to references (Russell (1983), and CACI (1983)).

4.3. NETWORK ll.5

NETWORK ll.5 is a simulation package written in SIMSCRIPT ll.5. It is intented to be a design that takes a user-specified computer system description and provide measures of hardware utilization, software executions, and contentions. It can be used to evaluate the ability of a proposed computer system to meet the required work load and to evaluate competing architectures. This simulation package is designed for the analysis of various computer systems, from a single processing element to a complex multiprocessor computer system, or a computer network. This package consists of three parts: NETIN, NETWORK, and NETPLOT. The queueing model of the computer system to be simulated is described to NETWORK ll.5 by a data structure consisting of processing elements, data transfer devices, data storage devices, modules, and files. Each of these entities have a series of attributes that have values supplied by the user.

NETWORK part reads in a data file describing the software and hardware parameters of the model. The input data file, that is provided by NETIN is a semi-English description of the computer system to be simulated. NETWORK builds and executes after it receives all required run time control parameters. The software of the queueing model of a computer system is presented to NETWORK ll.5 in the form of software modules. Each software module contains a description of processing elements allowed to execute the module, when it may run, what it is expected to do, and which other modules to start upon completion. The user can specify if copies of the same module may or may not execute concurrently on different processing elements.

Hardware devices are described as being one of three kinds: processing elements, data transfer devices, data storage devices (memories). Processing elements are specified by their instruction sets and cycle time. Data transfer devices such as buses, are characterized by their rates, protocols and connections. Data storage devices are characterized by size, access time, and access method (CACI (1985)).

NETWORK ll.5 provides a number of features mainly designed for computer systems simulation and computer networks. The most frequently used features are:

1. Statistical distributions.
2. Wild card(*): which is a feature for string matching. It can be utilised in message and file requirements as well as destination processors. For instance, if a message requirement is "XYZ*", then any message containing "XYZ" will satisfy the requirement.
3. Semaphore: these are global flags for control and synchronization between PEs. When a semaphore is set or cleared, the simulation time is reloaded. The statistical parameters such as mean, standard deviation, etc. are recorded at the end of simulation. Semaphore do not affect the accuracy of the simulation result since semaphore sets and clears do not consume any simulation time.
4. Message Echo: this is a suitable function for message acknowledgement simulation. The sent message can be specified as "echo" in order to get an acknowledgement.
5. Data Transfer Device Protocols: the package supports in addition to the default FCFS protocol other protocols including collision, token ring, slotted ring, and priority.

4.4. COMNET ll.5

COMNET ll.5 is a communication analysis package that is written in SIMSCRIPT ll.5 simulation language. It provides measures of computer network performance such as queueing delays, utilization, packet statistics and end-to-end blocking capabilities. These performance measures depend on the type of network traffic and resources.

COMNET ll.5 can simulate the routing of packets in computer networks and predict performance in terms of packet delay statistics, circuit group utilisation, etc. Data and integrated services digital networks can be modeled, including packet-switched networks with virtual-circuit or datagram operation (CACI (1988)).

The package provides a menu-driven interface through which you can describe network topology, traffic load, and routing algorithms. The network description proceeds very fast next to the built-in routing algorithms. Network description is saved in a database, and therefore, the changes are easily made. COMNET ll.5 reports provide summary of the performance measure statistics automatically. Description of a computer network include:

1. Network topology.
2. Network traffic descriptors for data messages (circuit, message or packet switched).
3. Routing algoritms, including predetermined alternate routing or adaptive shortest-path routing.
4. Simulation controls such as animation options, and run length.

COMNET ll.5 package provides error diagnosis capabilities in order to correct and resume the simulation.

5.   VERIFICATION AND VALIDATION

The issue of verification and validation is a vital one since this may affect the degree of accuracy and correctness of the developed simulation model. For small computer queueing models, the verification process is conducted by inspection, however, for larger models, some substantial analysis is required. A comparative "Walkthrough" of the model specification and the simulation program listing should be carried on. In some cases this inspection is enough to verify the simulation model if it is performed diligently. However, for other instances, verification by comparison with analytic models is made. The analytic verification provides a way to eliminate errors in at least part of the modeling process (Fishman (1978), Merle et al. (1987)). If the analytic verification is successful, then the remaining errors, if any exists, are either in extending assumptions from the analytic model to the simulation model, or in transforming the model specification to a model design.

In order to verify the transformation of a model description to a model design, a careful review is probably the best approach. The extension to simulation modeling from analytic modeling may require an added level of detail, queueing schemes and priorities. Parts of the model could be isolated and verified independently. In some cases parts of a closed system can be verified by isolating them and handling them as open system models, adding required event routines to generate arrivals and departures. We should often have an idea of the effect of modifying a specific assumption and check to see if the results change as expected. If, for example, we isolated part of a model and verified it as an open system, then we should expect its queueing delays to get reduced when merging it to become part of the closed system with a finite customer number. The effort spent in analytic verification can be minimized substantially if we plan from the start by providing parameters for the control extension and debugging, and verifying the model hierarchically. The main drawback of this approach is the tendency to bias the developments of the model toward verification rather than what it is meant to be. In addition to the general analytic models, specific previously derived analytic models can be useful in the verification procedure, especially if they have been validated.

The validation procedure is used to demonstrate that the simulation model is a good representation of the real system (i.e., it generates system behaviour with enough accuracy to satisfy analysis objectives). The accuracy degree depends on the objectives and perhaps the results obtained in the present iteration of the analysis procedure. The simulation model is usually developed to analyse a specific system in order to analyze a particular problem which in turn dictates the level of the required details. The simulation model is not necessarily required to be valid for all parts of the system over the entire spectrum of the systems behaviour; it just has to be valid for the requirements of the considered problem.

We can differentiate between two types of validation: for systems that can be measured and for systems that cannot be measured. In the first category, the system being modeled exists and therefore measurement can be made, and in such case, the goal is to predict a proposed change on the system. Validation in this case is based on the comparison between the model results and measurements. In the second category, the system being modeled exists only as a design and the simulation goal is to predict the performance of the design and optimize it. Validation of an existing sytem is basically a matter of comparison between the results of the existing system and that of the simulation model. If the two results agree, this means that the simulation of the modified system will produce valid estimates of the effects of the suggested change. Data used to validate the model in this case are collected when the work load characterization data is gathered. The process of collecting work load data and performance data for the work load conditions may seem to be an easy task, however, this can be difficult to perform. Measurement technique constraints can make it hard to match the beginning and end of the measurement period. This may result in incompleteness,inconsistency, and perhaps not required level of detail. One way to alleviate this problem is to develop a measurement model of the system. This model is derived from the analysis model description and it can provide a framework to evaluate and apply available measurement. In order to use this measurement model, it is necessary to validate it. The validation of this measurement model is performed by comparing aggregated work load data, with utilization and traffic measures (Welch (1983)). The simulation model's instrumentation should be modified to report the same performance indices as that of the system. Performance indices can be reported in the form of means and variances or distributions. Due to the limitation in reporting and statistical considerations, comparison between the measurements and model results is usually based on average values.

The work performed by the simulated system and the real one is the same in the context of statistics. The main difference between the measurement and the model values is due to the sampling variation and assumptions considered in the development of the model. When comparing, we should not expect the same percentage difference for all parameters. One type of subjective validation relies on the comparison of distributions. If we are able to get a report of the distribution of values, such as system resident time, from the measurement facilities, then we can make the model produce the corresponding distribution. Then the two distributions are compared by inspection. There are various statistical techniques used to compare measurement and simulation results and are explained in references (Heidelberger and Lavenberg (1984), Kiviat et al. (1969)).

When the differences between the measurements and simulation results are more than we like to accept, then we face a new task related to the model calibration. This problem is usually due to an invalid assumption that may have been made implicitly. Therefore, we have to check not just the explicit assumptions, but also a review of the system operation in order to see if we have left something out. In some cases the problem may be in the collection of data such as the incomplete measurement of overhead time due to the operating system.

When the system being modeled does not exist then the validation procedure is mainly done by review, i.e., the model is tested in terms of the system design and every assumption is examined and justified. The validity of the model can be tested by satisfying an expert audance. The review of the design team is based on experience rather on the factual knowledge. This review is basically a matter of comparing the model and system designs. This may be augmented by the knowledge of the system users. Workload representation is of a great concern especially when the goal of modeling is to predict the performance of the suggested design. When the goal of modeling is to compare alternative designs, we may use work load parameters that are appropriate to the case. In general, the more the model resembles the design, the more it is valid.

Validation is often merged with verification. If the measurements and model results are of close match, then simulation program is assumed to be a valid representation of the system and a valid implementation of the model design. The last step in validation is carried on when the system change is made or system design implemented and real performance can be compared. However, this does not happen usually, since the final realization may differ from the original design for reasons that have nothing to do with performance.

6. A CENTRAL SERVER MODEL OF A BATCH SYSTEM

The system to be studied in this section comprises a cpu and three disks. Two different types of fixed number of tasks execute in the system, alternating cpu and I/O (disks) activities. The cpu execution times for both classes of tasks (A and B) are assumed to be exponentially distributed with mean execution times of 20 μs and 15 μs for tasks A and B respectively. Class A contains $n_1 = 33$ tasks and class B contains $n_2 = 15$ tasks. Cpu requests of class B have pre-emptive priority over those of class A. The I/O requests of both classes are distributed randomly across all three I/O devices (disks) with the same mean disk service time of 40 μs. The I/O service times are assumed to be Erlang distributed with a standart deviation equaling 2/5 of the mean service time. Figure 5 shows the system to be studied.

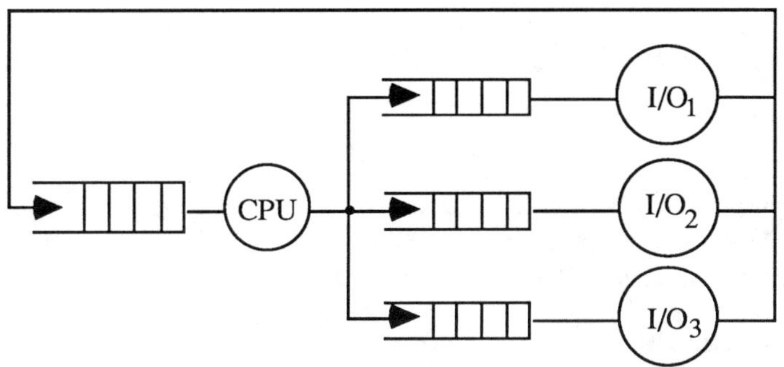

Figure 5. A Central Server Queueing System

In order to model this system using NETWORK ll.5, we considered the following entities for the model:

A. Statistical Distribution Functions:

1. EXP1: Exponential type with a mean of 20 μs to represent the distribution of class A cpu requests.
2. EXP2: Exponential type with mean 15 μs to represent the distribution of class B cpu requests.
3. ERL1: Exponential type with mean 10 μs to represent the number of repetitions of execution of each instruction.
4. ERLANG: Erlang type with mean of 40 μs and standart deviation of 16 μs to represent the service time of each I/O device.

B. Processing Elements (CPU)

The cycle time of the CPU is 0.1 μs and the following instructions are considered:

Read Instructions:

Read 1:Used in class A tasks to access Disk 1
Read 2:Used in class A tasks to access Disk 2
Read 3:Used in class A tasks to access Disk 3
Read 4:Used in class B tasks to access Disk 1
Read 5:Used in class B tasks to access Disk 2
Read 6:Used in class B tasks to access Disk 3

Write Instructions:

Write 1:Used in class A tasks to access Disk 1
Write 2:Used in class A tasks to access Disk 2
Write 3:Used in class A tasks to access Disk 3
Write 4:Used in class B tasks to access Disk 1

Write 5:Used in class B tasks to access Disk 2
Write 6:Used in class B tasks to access Disk 3

Processing Instructions:

PROC1:Used in class A tasks of EXP1 dist. type
PROC2:Used in class A tasks of EXP1 dist. type
PROC3:Used in class B tasks of EXP2 dist. type
PROC4:Used in class B tasks of EXP2 dist. type

C. Transfer Device (BUS):

Assumed to be ideal of zero transmission delay.

D. Storage Devices (Disks):

The storage devices: DISK 1, DISK 2 and DISK 3 are assumed to have word access times of ERLANG distribution types.

E. Macro Instructions:

Six macro instructions were used in the model:

MAC1:Used in class A test programs (test 1 to test 11).
        It has the instructions:PROC 1, READ 1, PROC 2, and WRITE 1.
MAC2:Used in class A test programs (test 12 to test 22).
        It has the instructions:PROC 1, READ 2, PROC 2, and WRITE 2.
MAC3:Used in class A test programs (test 23 to test 33).
        It has the instructions:PROC 1, READ 3, PROC 2, and WRITE 3.
MAC4:Used in class B test programs (test T1 to test T5).
        It has the instructions:PROC 3, READ 4, PROC 4, and WRITE 4.
MAC5:Used in class B test programs (test T6 to test T10).
        It has the instructions:PROC 3, READ 5, PROC 4, and WRITE 5.
MAC6:Used in class B test programs (test T11 to test T15).
        It has the instructions:PROC 3, READ 6, PROC 4, and WRITE 6.

All of these instructions in the aforementioned MACROS are assumed to have an ERL1 distribution.

F. Software Modules (test programs):

Two classes of test programs A and B are to be executed by the CPU. Class A contains 33 test programs of the same priority:

Test 1 to Test 11 : Access DISK 1
Test 12 to Test 22 : Access DISK 2

Test 23 to Test 33 : Access DISK 3

Class B contains 15 test programs of the same priority as shown below:

T 1 to T 5 : Access DISK 1
T 6 to T 10 : Access DISK 2
T 11 to T 15 : Access DISK 3

Remember that class A tasks (Test 1 to Test 33) have priority keys equal to 0, where tasks (T1 to T15) have priority keys equal to 1, i.e., tasks of class B have higher priority than those of class A.

G. Results:

The required performance indices are found as explained below:

1. The time a task spends in the system differs from one to another. The average execution times of both classes are:

Class A: Average execution time $= 11973.008$ μs (non-preemptive case)
$\qquad\qquad\qquad\qquad\qquad\quad = 21857.74$ μs (preemptive case)
Class B:Average execution time $= 5616.502$ μs (non-preemptive)
$\qquad\qquad\qquad\qquad\qquad\quad = 9022.27$ μs (preemptive)
Total average $= (11973.008 + 5616.502 * 15)/48 = 9986.6$ μs (non-preemptive)
Total average $= (21857.740 + 9022.270 * 15)/48 = 17846.66 \mu s$ (preemptive)

These results are expected since class B tasks have priority over those of class A.

2. Utilization:

CPU   Utilization $= 47.934\%$ (non-preemptive)
$\qquad\qquad\qquad\qquad = 72.131\%$ (preemptive)
DISK 1 Utilization $= 10.449\%$ (non-preemptive)
$\qquad\qquad\qquad\qquad = 29.670\%$ (preemptive)
DISK 2 Utilization $= 24.149\%$ (non-preemptive)
$\qquad\qquad\qquad\qquad = 17.928\%$ (preemptive)
DISK 3 Utilization $= 12.037\%$ (non-preemptive)
$\qquad\qquad\qquad\qquad = 22.610\%$ (preemptive)
BUS   Utilization $= 46.635\%$ (non-preemptive)
$\qquad\qquad\qquad\qquad = 70.208\%$ (preemptive)

3. Mean Queue Length $= 9.597$ (nonpreemptive)
$\qquad\qquad\qquad\qquad\quad = 15.557$ (preemptive)

4. Mean Busy Periods (Average Usage Times)

DISK 1 Average usage time $= 68.611$ μs (non-preemptive)
$\qquad\qquad\qquad\qquad\qquad = 69.000$ μs (preemptive)
DISK 2 Average usage time $= 71.681$ μs (non-preemptive)

$$= 70.554 \ \mu s \ \text{(preemptive)}$$

DISK 3 Average usage time $= 69.335 \ \mu s$ (non-preemptive)
$$= 71.551 \ \mu s \ \text{(preemptive)}$$

BUS Average usage time $= 70.361 \ \mu s$ (non-preemptive)
$$= 70.201 \ \mu s \ \text{(preemptive)}$$

5. The number of complete facility operations(i.e., difference between the release count and the preemp count) are listed in Table 1. It is observed from these results that class B tasks have less average execution time than class A, and this is justified since class B tasks have priority over those of class A in both the preemptive and non-preemptive cases. Moreover, there are more operations completed in the preemptive case and hence the average execution times of both classes in the preemptive priority case are more than those in the non-preemptive case. A final comment on the simulation results is that the CPU and BUS are more heavily utilized in the preemptive case as compared to the non-preemptive case. Copies of the simulation program listing and results can be obtained directly from the author.

Table 1. Number of Operations for Non-preemptive and Preemptive Cases

	Non-Preemptive	Preemptive
CPU	6228	10001
DISK 1	1523	4300
DISK 2	3369	2541
DISK 3	1736	3160

7. A TIME-SHARED COMPUTER MODEL

In this example, we consider a model of a time-shared computer system that has a single CPU, disk drive, tape drive and 15 terminals. The user of each terminal thinks for an amount of time which is an exponentially distributed random variable with a mean of 25 seconds, after which he sends a job to the computer. The sent job joins a FIFO queue at the CPU. The service times of jobs at the CPU are exponential random variables with a mean of 0.8 second. The job, upon leaving the CPU, is either finished with a probability of 0.2 independent of the system state and returns to its terminal for another think time, or it requires data from the disk drive with probability 0.72, or needs some data available on the tape with probability 0.08. If a job leaving the CPU is sent to the disk drive, it may have to join a FIFO queue until the disk is free. The service time at the disk drive is an exponential random variable with a mean of 1.39 sec. When finishing disk service, the job queues up again at the CPU. A job leaving the CPU bound for the tape drive has an experience similar to a disk job, except that a service time at the tape drive is an exponential random variable with a mean of 12.5 sec. Think times and service times are independent and all terminals are in the think state at time zero.

The system will be simulated for 1000 job completion and statistics will be gathered. Figure 6 shows a simplified diagram of the system to describe the performance of the system; in particular, the utilization of all three devices, average number of jobs in the queue, mean delay in the queue, and maximum and mean response time of a job.

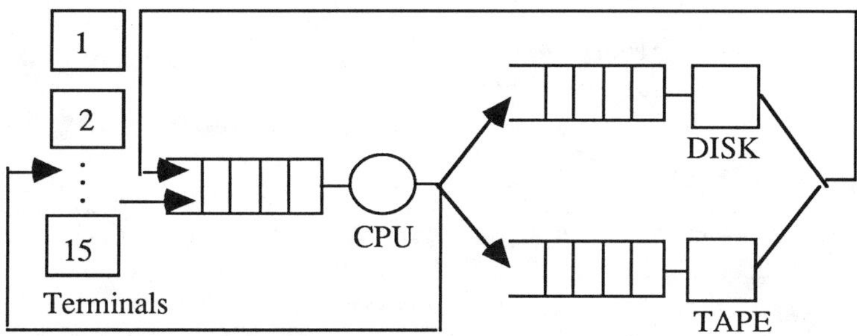

Figure 6. A Time-Shared Computer Model

The model for the system is developed using the SIMSCRIPT ll.5 simulation language. The variables used in the simulation program are shown in Table 2.

Table 2. Variables Used in the Model for the Time-Shared System

Variable	Definition
JB.TERMINAL	Attribute of process JOB used to represent the terminal corresponding to a particular job.
AVG.NUMBER.IN.QUEUE	The time-average number of jobs in queue.
MAX.RESPONSE.TIME	Maximum response time of a job.
MAX.TERMINALS	Maximum number of terminals.
IN.NUM.TERMINALS	Increment the number of terminals.
MEAN.SERVICE.TIME	Mean service time.
NUM.TERMINALS	Number of terminals for a particular simulation
RESPONSE.TIME	Response time of a particular job.
SERVICE.TIME	Service time of a particular job.
START.TIME	Time that a job leaves its terminal.
UTIL.CPU	Utilization of the CPU.
MEAN.THINK.TIME	Mean think time.
MEAN.RESPONSE.TIME	Mean response time.

The job will be modeled from the instant it leaves the terminal by a process called JOB. A process called TERMINAL is used to model the terminal operator thinking before sending a job to the CPU to be executed (processed). The variables used in the model are all global except SERVICE.TIME and START.TIME which are local variables in process routine JOB. The listings of the PREAMBLE, MAIN, routines read data, process terminal, process job and report are all listed in Figures 7,8,9,10,11 and 12 respectively, and the simulation results are listed in Figure 13.

```
preamble

  Processes include Terminal
     Every job has a jb.terminal
     define jb.terminal as an integer variable

  Resources include CPU, DISK, TAPE

     Define mean.service.time,mean.think.time,DELAY.IN.TAPE.QUE
        DELAY.IN.DISK.QUE,DELAY.IN.CPU.QUE and
        response.time as real variables

     Define inc.num.terminals,max.terminals,min.terminals
        num.jobs.desired,num.jobs.completed, and num.terminals
        as integer variables

     Define .seconds to mean units

     Tally max.response.time as the maximum and
        mean.response.time as the mean of response .time

  Accumulate avg.num.in.CPU.que as the average of n.q.CPU
  Accumulate avg.num.in.disk.que as the average of n.q.disk
  Accumulate AVG.NUM.IN.TAPE.QUE AS THE AVERAGE OF N.Q.TAPE
  Accumulate AVG.DELAY.IN.TAPE.QUE
    AS THE AVERAGE OF DELAY.IN.TAPE.QUE
  Accumulate AVG.DELAY.IN.CPU.QUE
    AS THE AVERAGE OF DELAY.IN.CPU.QUE
  Accumulate AVG.DELAY.IN.DISK.QUE
    AS THE AVERAGE OF DELAY.IN.DISK.QUE
  Accumulate UTIL.DISK AS THE AVERAGE OF N.X.DISK
  Accumulate UTIL.TAPE AS THE AVERAGE OF N.X.TAPE
  Accumulate util.CPU as the average of N.X.CPU

end
```
Figure 7. Listing of the PREAMBLE of the Time-Shared System Model

```
main
   call read.data
   for num.terminals=min.terminals to max.terminals
   by inc.num.terminals do
   call initialize
   start simulation
   loop
end
```

Figure 8. Listing of the MAIN Program


```
routine read.data
   create every cpu(1)
   let u.cpu(1)=1
   create every disk(1)
   let u.disk(1)=1
   create every tape(1)
   let u.tape(1)=1

      read min.terminals,max.terminals,inc.num.terminals
        and mean.think.time
      read mean.service.time and num.jobs.desired
      print 6 lines thus

Time-Shared Computer Model -- Input Parameters

skip 3 lines
print 7 lines with mean.think.time,mean.service.time
and num.jobs.desire
   THUS

Think times are exponential with mean ***.* seconds

service times are exponential with mean ***.* seconds

Number of jobs processed *****

start new page
print 6 lines thus

Simulation results (All times are in seconds)

skip 3 lines

end
```

Figure 9. Listing of the READ.DATA Routine

```
process terminal
until num.jobs.completed >= num.jobs.desired
do /
  wait exponential.f(mean.think.time,1) .seconds
  create a job
  let jb.terminal(job)=terminal
  activate this job now
  suspend
  add 1 to num.jobs.completed
  if num.jobs.completed=num.jobs.desired
    call report
  always
  loop
end
```

Figure 10. Listing of the PROCESS.TERMINAL Routine

```
Process Job
  define start.time AND prob as real variables
  let.start.time=time.v
  let prob=1
  while prob>=0.2
  do
    let enter.cpu.que=time.v
    request 1 cpu(1)
    delay.in.cpu.que=time.v-enter.cpu.que
    WORK(exponential.f(mean.service.time,2)) .seconds
    relinquish 1 cpu(1)
    LET prob=random.f(1)
    if prob <= .92
       let enter.disk.que=time.v
       request 1 disk(1)
       delay.in.disk.que=time.v-enter.disk.que
       work(exponential.f(1.39,1)) .seconds
       relinquish 1 disk(1)
    else
       let enter.tape.que=time.v
       request 1 tape(1)
       let delay.in.tape.que=time.v-enter.tape.que
       work(exponential.f(12.5,1)) .seconds
       relinquish 1 tape(1)
```

```
    always
  loop
  let response.time=time.v-start.time
  reactivate the terminal called jb.terminal(job) now
end
```

Figure 11. Listing of the PROCESS.JOB Routine

```
routine report
  print 14 lines with num.terminals,mean.response.time
    max.response.time
  avg.num.IN.CPU.QUE(1), util.cpu(1), AVG.DELAY.IN.CPU.QUE,
  AVG.NUM.IN.DISK.QUE(1), UTIL.DISK(1), AVG.DELAY.IN.DISK.QUE,
  AVG.NUM.IN.TAPE.QUE(1), UTIL.TAPE(1), AVG.DELAY.IN.TAPE.QUE
  thus

Number of Terminals:  ***

      Mean Response Time:*****.***    Max.Response Time:*****.***
      Time Average Of        Device  Average Delay
        # Jobs In Que     Utilization     In Que
CPU         ***.***           *.***       ***.***
DISK        ***.***           *.***       ***.***
TAPE        ***.***           *.***       ***.***
end
```

Figure 12. Listing of the REPORT Routine

```
Think times are exponential with mean 25.0 seconds
service times are exponential with mean .8 seconds
Number of jobs processed 1000

Simulation results (All times are in seconds)

Number of Terminals:  15

      Mean Response Time:  70.519    Max.Response Time:  570.66
      Time Average Of        Device  Average Delay
        # Jobs In Que     Utilization     In Que
CPU            .876            .614          .841
DISK          6.563           .973         9.207
TAPE          1.339           .670        15.654
```

Figure 13. Simulation Results of the Time-Shared Computer System Model

8. A MULTIPROCESSOR COMPUTER SYSTEM MODEL

This section is devoted to the discussion of the simulation of a multiprocessor computer system. The system is a time-shared common bus multiprocessor system that consists of four processors with their individual local memories, a common bus, arbiter and a common memory that is shared by all processors in the system. Each of the processing elements can access the common memory via a global bus (common) CB. Programs executed by each processor are generally not independent of each other, therefore, interprocessor communications are required.The system is shown in Figure 14.

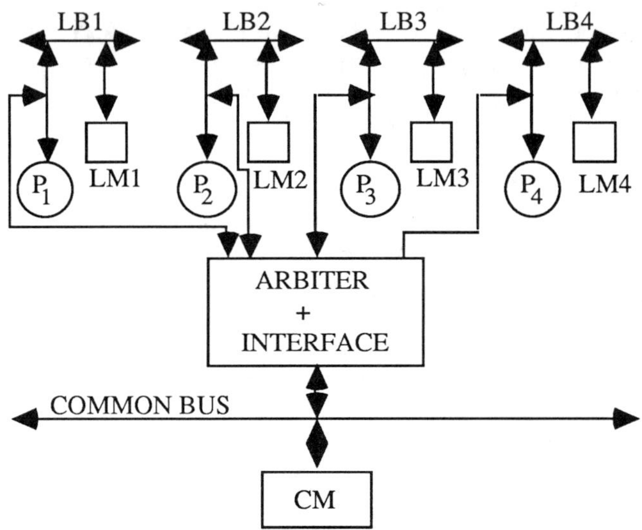

Figure 14. The Multiprocessor Computer System

Below is a description of the characteristics of the blocks of the system:

1. The Arbiter arbitrates processors' requests to the common BUS (CB) where the common memory is connected.

2. Local memories LM_i and Common Memory CM are all single-ported memory devices.

3. Size of LM_i is 50 kbyte
   Size of CM is 1 Mbyte

4. Access times: for LM_i $T_{acc} = 7$ ηs
   for CM $T_{acc} = 50$ ηs

5. Processors: Cycle time $= 0.1 \mu s (100$ $\eta s)$
   Word size $= 16$ bits

6. Common bus and Arbiter:

Protocols: The data transfer protocols to be considered are: FCFS, Token Ring and priority schemes. The CB is assumed to have a speed of $0.1\mu sec$/word.

7. Load characteristics:

No benchmark programs are used for this model. The Te_i's (execution times) were measured using a logic analyser and a single processor system. Every processor runs for a duration of time Te_i, then it submits a request to access data from the common memory. Depending on the status of the CM, arbitration scheme and electronics characteristics of the arbiter, the processor waits for a Tw_i time until it gets permit to use the common memory. Then, it accesses the CM for T_s (access time) units. This sequence is repeated until the task is completed.

In order to model the system using NETWORK 11.5 simulation package the benchmark programs were organised as follows:

1. Benchmark 1

There are n repetitions for the first four instructions: Te_1, READ, Te_2, and WRITE. We defined a macro instruction MAC1 which contains the above instructions to be a single instruction listed in the program instruction list and n is chosen to be 2000. Table 3 shows the parameters of benchmark program 1.

Table 3. Execution Times of Benchmark Program 1

Te_i	Time in ηs
Te_1	4.0
READ from CM	0.5
Te_2	2.0
WRITE to CM	0.5
Te_3	4.0
READ from CM	0.5
.	.
.	.
.	.
Te_n	2.0

2. Benchmark 2

Here, there are n repetitions for the first six instructions: Te_1, READ, Te_2, WRITE, Te_3, and READ. We defined a macro instruction MAC2 which contains the above instructions and n is chosen to be 1000. Table 4 shows the parameters of benchmark program 2.

Table 4. Execution Times of Benchmark Program 2

Te_i	Time in ηs
Te_1	3.5
READ from CM	0.5
Te_2	0.8
WRITE to CM	0.5
Te_3	3.5
READ from CM	0.5
.	.
.	.
.	.
Te_n	0.8

Each one of the four processors is modeled as a processing element executing the specified benchmark program. The buses and memory devices are modeled as transfer devices and storage devices, respectively, and using the hardware characteristics given in the example. The access times of memories are deterministic in this example and the sizes are 70 kbyte and 1 Mbyte for local and common memory respectively. Figure 15 shows a simplified queueing model diagram for the system.

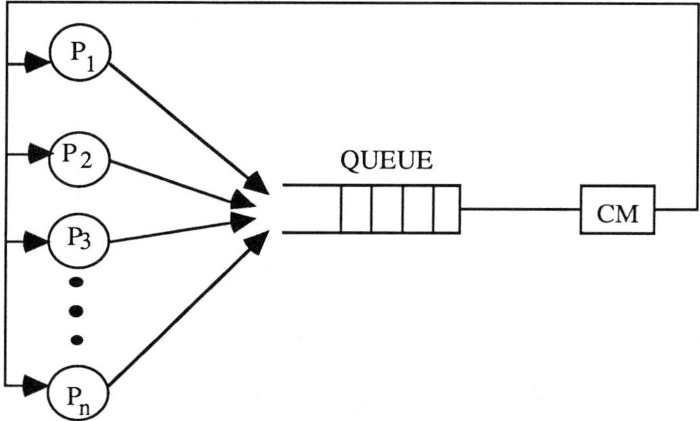

Figure 15. The Multiprocessor Queueing Model

The model was developed for the two benchmark programs by considering three different arbitration disciplines for the common memory access. These arbitration schemes are FCFS, Token Ring, and priority.

From the simulation runs we were able to find the relationship between speedup, and the number of processors in the system. The speedup is defined as

$$Sp_n = \frac{\text{Time required by one processor to execute a task}}{\text{Time required by n processors to execute the same task}}$$

To apply this definition here, we let

$$Sp = \frac{\text{Time required by one processor to execute a task program N times}}{\text{Time required by N processors to execute the task program once}}.$$

The Sp versus the number of processors in the system relation is plotted in Figures 16,17 for benchmark programs 1 and 2 respectively.

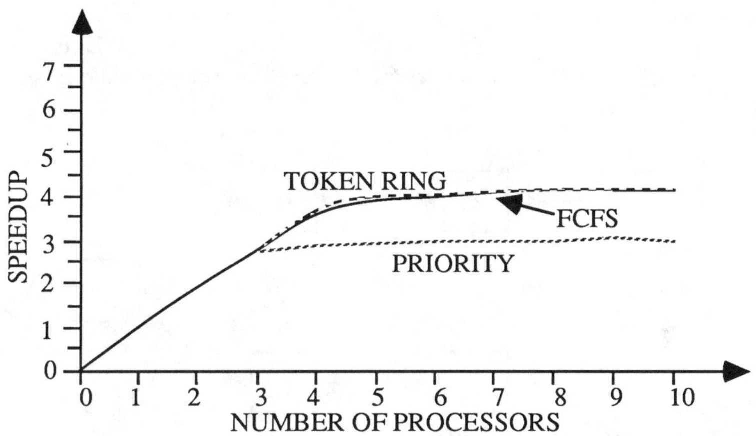

Figure 16. Speedup vs Number of Processors in the System for Benchmark Program 1

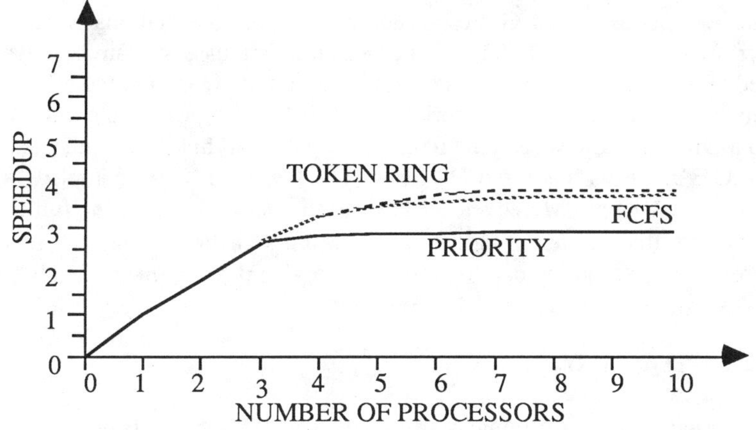

Figure 17. Speedup vs Number of Processors in the System for Benchmark Program 2

The performance of the system increases proportionally as the increase in the number of processors in the system until some critical value where you do not get any further speedup improvement. This is due to the bus traffic which usually depends on the nature of the application program, the arbitration discipline, and other minor facts.

It is observed from simulation results that benchmark for 1, the FCFS and Token Ring provides better speedup than the priority scheme especially when the number of processors is high. The same observation applies for benchmark 2 except for a minor difference.

9.   A COMPUTER NETWORK MODEL

The example described in this section is a simulation model of a long-haul computer network. The network consists of seven nodes (computers) A,B,C,D,E, F and G. These nodes represent branch offices of some establishment.   The network is illustrated in Figure 18.

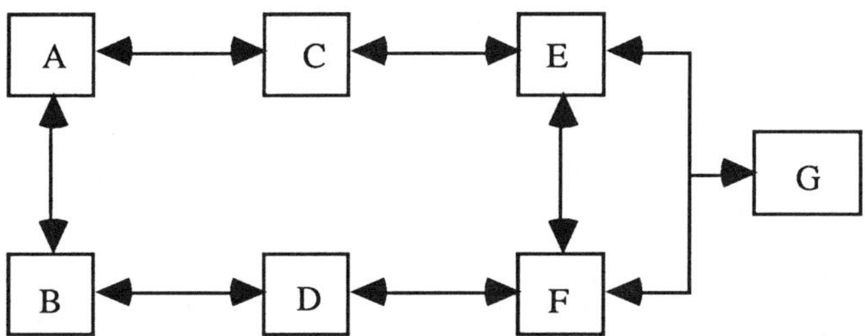

Figure 18. A Long-Haul Computer Network

The two nodes A and G are considered as data base stations which are accessed by other nodes. Nodes C,D and B can access the data base station A while station G is accessed by nodes E and F. Moreover, data base stations A and G regularly exchange long files between each other. The network has two different types of data links, namely high-speed data link and low-speed data link. The high-speed links are connected between the nodes A,C,E and F with a rate of 32 Kbps. The low-speed links complete the remaining connections at 4 Kbps. Moreover, all links in the network are of the full-duplex type.

There are three different categories of traffic signalled among the seven branches, these are: data base query, data base query reply, and file transfer. Each kind of traffic has its own statistics as illustrated below.

a. Data-Base Query Traffic

This kind of traffic originates from secondary branches (B,C,D) and (E,F) to the primary data base stations A and G respectively. The query traffic occur exponentially with mean 2.5 sec. (i.e., interarrival time is exponential), the message size distribution is constant with 256 bits (32 bytes).

b. Data-Base Query Reply

This traffic originates from the data base stations A and G to their local branches as a reply on their queries. The interarrival time distribution is exponential with mean 0.5 sec.. The query reply message size is constant with 1024 bits (128 bytes).

c. File Transfer Traffic

The origin and destination of this traffic is the data base station nodes A and G. The interarrival time distribution is exponential with a mean of 3 sec. (i.e., time between two consecutive files (messages) is 3 seconds on the average). Message size distribution is normal with mean and standart deviation of 600 and 200 bytes respectively.

The considered network is of a packet-switching type. Therefore, the alternative routing features may be exploited in such a way that packets can be routed via different paths to improve the performance of the network. Static routing algorithm is used here and the routing criterion is the minimum queue size of links. Table 5 shows the routing path for each node. Each cell of the table contains destination nodes that can be chosen by an origin node. The destination nodes in each cell are arranged in such a way that the first one is the first choice, and so forth.

Table 5. Routing Table of the Long-Haul Computer Network

From \ To	A	B	C	D	E	F	G
A	-	B	C	C B	C B	C B	C B
B	A	-	A D	D	A D	A D	A D
C	A	A D	-	D	E	E D	E D
D	C B	B	C	-	C F	F	C F
E	C F	C F	-	C F	-	F	G
F	E D	E D	E D	D	E	-	G
G	E F	A D	E F	E F	E	F	-

The COMNET ll.5 simulation package is used to model the network. Therefore, additional parameters that complete the specification of the network model are required. These are explained below:

Packet Buffet Size (Maximum Queue Length):

This defines the total amount of space available at a packet switch (node) for packets that are waiting for an outgoing data link. This parameter will be fixed during simulation; however, it will be varied to study its effect on the characteristics of the network.

Class of Service (COS):

The COS can be used to define categories of message with the same origin and destination. For example, node A originates two different message categories: file transfer and query reply. We define three types of COS : FIL, RPL and QER. These classes represent file transfer, query reply, and query traffic respectively.

Packet Length:

The length of the packet is assumed to be fixed of 128 bytes of information plus three overhead bytes. The overhead bytes accompany the information bytes from the originating node to the destination. These are used mainly for routing. There are three points to be mentioned regarding the simulation of this network. First, the simulation is performed for 15 minutes. Second, node "A" was arbitrated so as to highlight the simulation results. Third, the discussion is concentrated on the high-speed link "A-C". The performance of the network is examined and the behaviour of the queueing parameters are investigated as well. We have conducted two parts of simulation:

A. The maximum node queue length and its effect on the characteristics of the whole network.

B. The effect of the queue length of link "A-C".

A. Maximum Queue Length (Buffer Size)

There are various papers that study the effect of finite buffer size allocated to nodes (switches) and provide design curves relating to the buffer overflow (message/packet blocked) to buffer size and message/packet statistics. Here, we simply describe the effect of the maximum queue length (buffer size) of nodes on the message/packet statistics.

COMNET ll.5 permits the user to define a buffer size that can be allocated to each node. This buffer is used for arriving packets as shown in Figure 19. In fact, the queueing system of this network is of type G/D/1/K. In each node, only one processor exists, therefore, it is a single server queue. The interarrival time is general since each node has two interarrival times; for instance, node A originates exponentially 1/3 messages per second and at the same time the arrival rate from outside (nodes C and B) is distributed normally with mean 1/0.25 messages per second. In addition, packets may be arbitrary routed from neighbouring nodes.

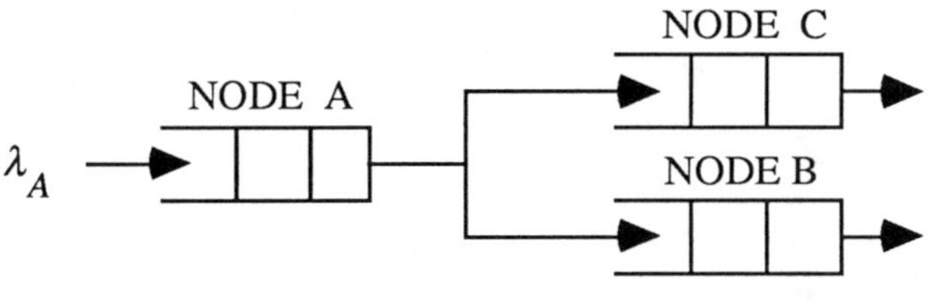

λ_A Arriving Rate Pkt(Msg)/second

Figure 19. The Queue Length Configuration for Node A and its Neighbors

Maximum queue length at each node affects the network characteristics, especially message per packet delay time, number of packets processed and blocked (buffer overflow) and packet queue time of a link. We varied the maximum queue length at each node and ran the simulation for 15 minutes for each case. Table 6 shows the simulation results obtained from eight experiments.

Table 6. Overall Message and Packet Delay

A-C Queue Length (Byte)	Max.Msg.delay (ms)	Avg.Msg.Delay (ms)	Avg.Pkt.Delay (ms)
160	60011	2000	3888
256	55011	1480	2675
320	30011	710	1261
600	10089	310	447
768	10012	240	314
896	6015	180	233
1024	5088	160	211

The overall average message delay, maximum message delay and packet average delay are tabulated as well. It is obvious that if the maximum queue length allocated for nodes increases, then the average/maximum message delay increases. Moreover, the packet average delay for the whole network has the same characteristic. Note that when the maximum length (buffer size) is very small (128 bytes), the above statement is not true. Figures 20,21,22 depict the behaviour of the network message/packet delay versus the maximum queue length.

In addition to the total number of blocked packets (buffer overflow) at a node, the maximum queue length affects the packet queueing time of a link. From Table 7 the reader can see that the ratio of the number of blocked packets (Nb) over the number of processed packets (Np) at node "A" increases as the maximum queue length decreases.

When the maximum allowable queue length at node A increases, consequently, the queue length of the link "A-C" will also increase. Therefore, the number of packets waiting in the "A-C" queue will grow as well. Moreover, packet wait time at node A will also increase. This is clearly shown in Table 8 and Figure 23.

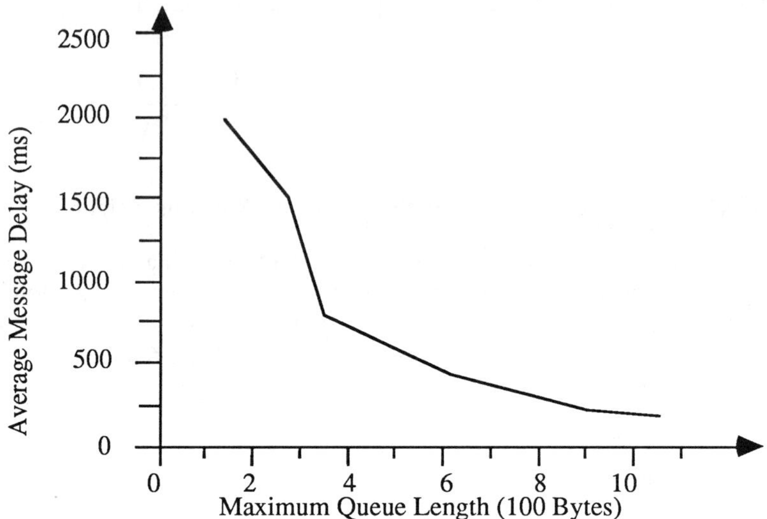

Figure 20. Queue Length vs. Overall Average Message Delay

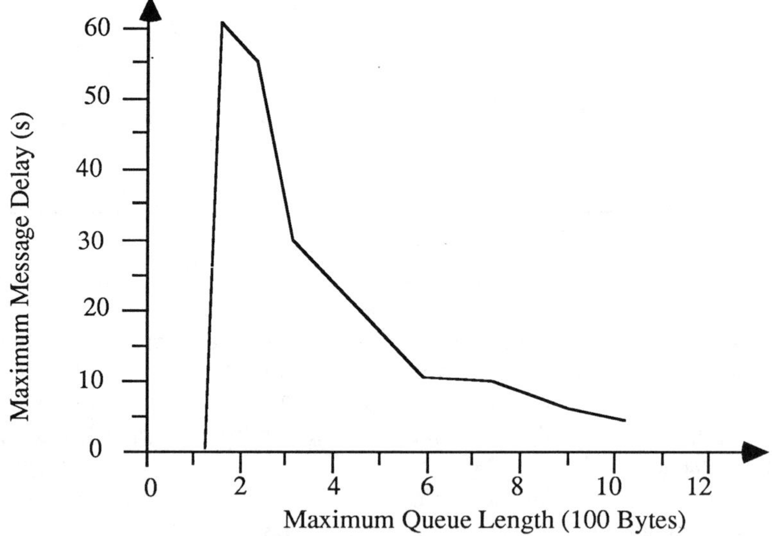

Figure 21. Queue Length vs. Overall Maximum Message Delay

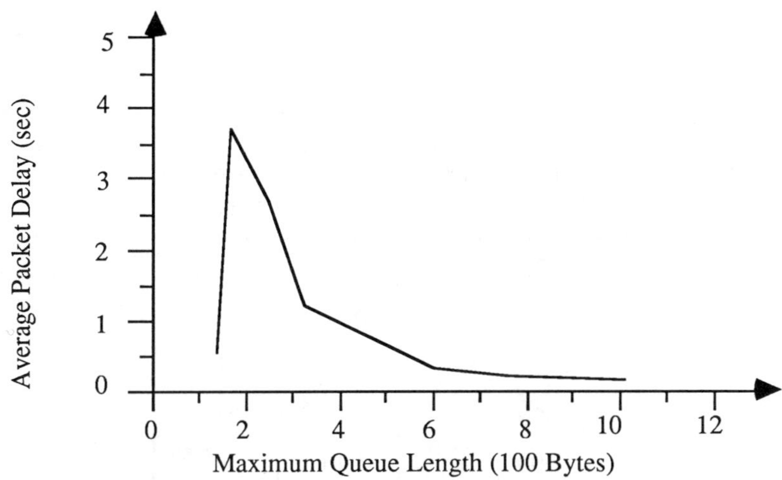

Figure 22. Queue Length vs. Overall Average Packet Delay

Table 7. Maximum Queue Length, Number of Processed and Blocked Packets Relationship at Node "A"

Max Queue Length (Byte)	No.of Processed Packets (Np)	Number of Blocked Packets (Np)	Nb/Np -
128	4887	1243	0.2543
160	4532	1137	0.2509
256	4095	798	0.194
320	3819	379	0.09925
600	3814	94	0.0246
768	3773	46	0.01219
896	3744	14	0.00373
1024	3733	3	0.00080

Table 8. Packet Queueing Time of Link "A-C"

Max. Queue Length (Bytes)	Pack. Queue Time	
	Avg (ms)	Max (ms)
128	0.21	30.89
160	0.25	31.53
256	3.12	32.50
320	3.14	34.50
600	3.46	48.25
768	4.26	66.75
896	5.34	89.00
1024	5.65	98.00

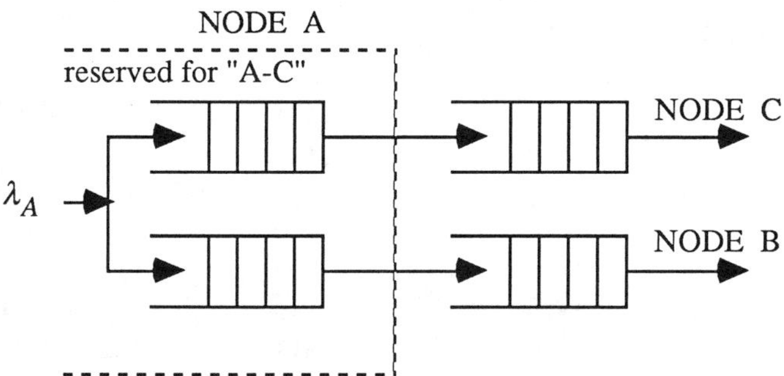

Figure 23. A-C Queue Configuration

B. Queue Length of Link "A-C"

In part A, link "A-C", like the other links in the network, was allowed to use a maximum queue length equal to the maximum queue length at node A. In this section we allocate a variable length of a queue dedicated for the link "A-C". This space is reserved from the available space at the origin node A. Similar to the other nodes, we assume node "A" has a maximum buffer size of 1024 bytes, and the reserved queue length for link "A-C" is varied from 160 to 1024 bytes. The simulation time is 15 minutes as in part A.

The simulation results are tabulated in Tables 9,10,11. It is observed from Table 9 that varying the maximum queue length of link "A-C" does not have a great effect, in general, on the performance of the network. Figures 24,25, and 26 show the plots of "A-C" queue length versus the overall packet delay, overall average message delay and overall maximum message delay respectively.

The two data base station nodes A and G are the most efficient nodes since all the other branches communicate with them. Simulation results show that they always process the highest number of packets.

Link "A-C" and "G-E" have the busiest traffic load while links "E-F" and "F-D" are the least efficient links which are idle most of the time. This is due to the unfair distribution in the routing table. Thus, this can be avoided by redistribution of the routing paths if needed.

Table 9. Overall Message and Packet Delay

A-C Queue Length (Byte)	Max.Msg.delay (ms)	Avg.Msg.Delay (ms)	Avg.Pkt.Delay (ms)
160	6042	200	325
256	6017	200	285
320	5084	180	230
600	5089	170	219
768	5089	170	194
896	5089	170	166
1024	5089	170	153

Table 10. Packet Blocked and Processed

A-C Queue Length (Byte)	Np	Nb	Nb/Np
160	3751	21	0.00559
256	3750	20	0.00533
320	3739	9	0.00240
600	3738	8	0.00186
768	3746	7	0.00186
896	3739	6	0.00160
1024	3733	5	0.00080

Table 11. Maximum and Average Queueing Time of Link "A-C"

"A-C" Queue Length (Bytes)	Pack. Queue Time	
	Avg (ms)	Max (ms)
160	0.18	28.89
256	1.24	30.50
320	2.41	45.00
600	5.89	111.35
768	13.55	156.00
896	16.17	198.10
1024	21.30	220.50

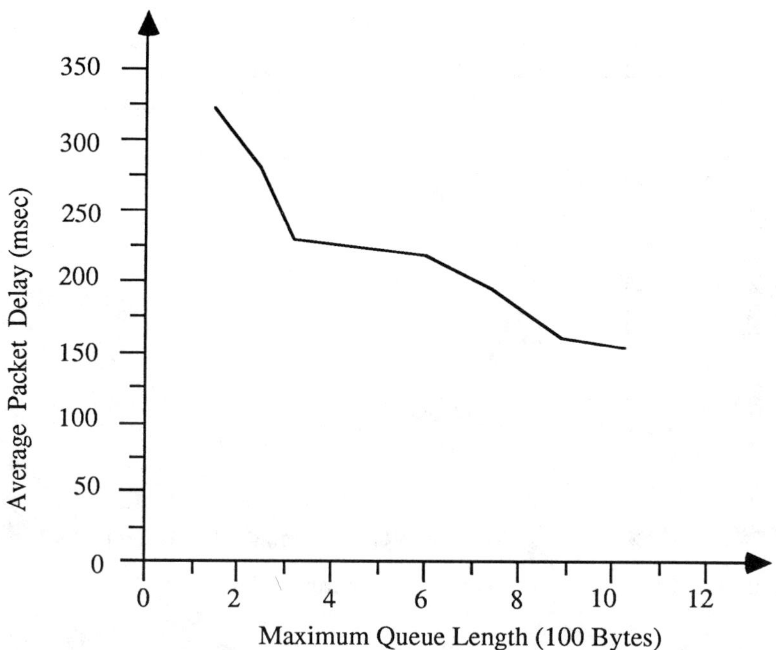

Figure 24. A-C Queue Length vs Overall Average Packet Delay

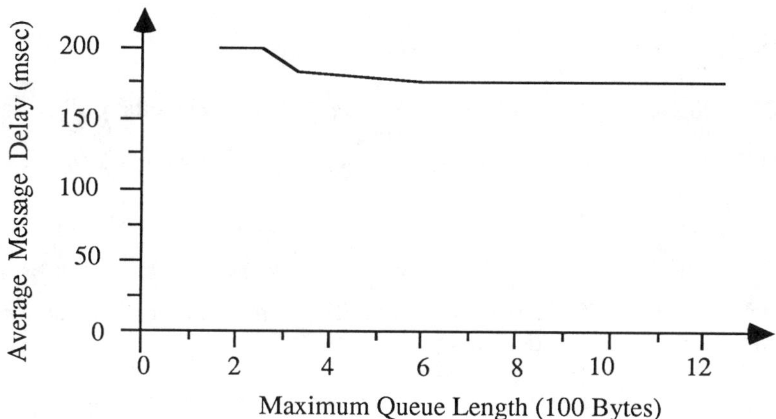

Figure 25. A-C Queue Length vs Overall Average Message Delay

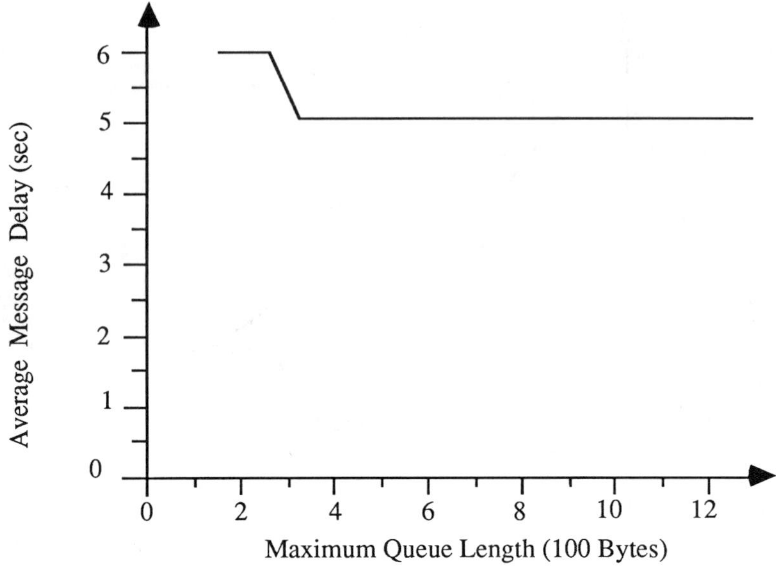

Figure 26. A-C Queue Length vs Overall Maximum Message Delay

10. REFERENCES

1. CACI. 1988. *COMNET ll.5 User's Manual.* CACI, La Jolla, California.

2. CACI. 1985. *NETWORK ll.5 User's Manual.* CACI, La Jolla, California.

3. CACI. 1983. *SIMSCRIPT ll.5 Programming Language.* CACI, La Jolla, California.

4. Fishman,G.S. 1978. *Principles of Discrete Event Simulation.* McGraw-Hill, New York.

5. Gordon, G. 1975. *The Application of GPSS V to Discrete Systems Simulation.* Prentice-Hall, Englewoods Cliffs, New Jersey.

6. Heidelberger,P. and Lavenberg,S. 1984. Computer Performance Evaluation Methodology. *IEEE Trans. on Computers* **C-33**, 108-133.

7. Kiviat, P.J., Villanueva, R., and Markowitz, H. 1969. *The SIMSCRIPT ll Programming Language.* Prentice-Hall, Englewood Cliffs, New Jersey.

8. Law, A.A. and Kelton, W.D. 1982. *Simulation Modeling and Analysis*. McGraw-Hill, New York.

9. MacDougall, M.H. 1987. *Simulating Computer Systems*. MIT Press, Cambridge, Massachusettes.

10. Merle,D.,Potier,D.,and Veran,M. 1987. A Tool for Computer System Performance Analysis. In *Performance of Computer Installations* , D.Ferrari(Ed.), North-Holland, Amsterdam, 195-213.

11. Motorola. 1983. *MC68000 16 Bit Microprocessor*. Motorola Inc., Austin,Texas.

12. Obaidat, M.S. 1989. Performance Evaluation of the MPS Multiprocessor Computer System. *Proceedings of the IEEE Compeuro-89*, Hamburg, Germany.

13. O'Donovan, T.M. 1979. *GPSS Simulation Made Simple*. McGraw-Hill, New York.

14. Russell, E.C. 1983. Building Simulation Models with SIMSCRIPT ll.5. CACI, La Jolla, California.

15. Sauer, C.H. and Mac Navir, E.A. 1983. *Simulation of Computer Communication Systems*. Prentice-Hall, Englewood Cliffs, New Jersey.

16. Schriber, T.J. 1974. *Simulation Using GPSS*. McGraw-Hill, New York.

17. Welch, P.D. 1983. The Statistical Analysis of Simulation Results in the Computer Performance Modeling Handbook. S.S. Lavenberg (Ed.), Academic Press, New York, 268-328.

Chapter 8

A PERSPECTIVE
ON
ACADEMIC COMPUTER NETWORKING

Ali Rıza Kaylan
Boğaziçi University
Department of Industrial Engineering
Bebek, Istanbul, Turkey

1. POTENTIAL BENEFITS OF CAMPUS NETWORKING

Two major factors which have a profound impact on the computing philosophy of the university campuses are the evolution of the network technology and the dramatic decrease in the cost of computing power. It is rather obsolete to have one large computer to meet the computational needs of the entire university. Obviously, there is a growing tendency to install personal systems scattered throughout the campus. However, the intention is not to generate a picture of disconnected islands of computer resources over the campus. Computer networks link mainframes, minis, and personal computers and peripherals into an efficient and cost-effective academic computing environment.

There are a number of benefits gained in arranging computer systems into networks. In this way, computer users take advantage of the university's total computing capabilities by sharing expensive resources. The reasons for using personal workstations as a scientist's workbench include increased productivity, easier personalization, ease and low cost of attaching specialized equipment. Mainframe computers, on the other hand, offer capabilities and resources on a larger scale such as high-speed computation and extremely large-scale storage.

Networking environment blends seamlessly the relative advantages of different levels of computing, namely workstations and large systems. The personal workstation provides ready access and gives immediate response. One can personalize the viewing screen like arranging work on a desk. Bit-map displays allow for color, graphics, image, and motion. Larger systems can better manage jobs such as those requiring high-speed calculations, shared program libraries, large memories, central data bases, computer conferencing facilities, network connections, and batch operations. Through the network, small stand-alone systems can access the greater power of large systems and their available resources while large systems can download applications most suitable for personal workstations. There are successful experiences reported combining workstations and supercomputers into a coherent structure to increase research productivity.

152

Fast computational work which is handled on the supercomputer side leads into quantitative productivity. Better visualization of scientific results ,however, are presented on the high resolution, interactive graphical displays of workstations reflecting qualitative productivity (Rounds (1989)).

The second principal benefit of networking is related to system reliability and availability. When computer systems are isolated from each other, a machine failure possibly results in a complete loss of computing power for the local users even though there may be substantial computing capacity available at the other sites of the campus. Networks enable those users who are temporarily out of any local resources to access other systems and fulfill their computing needs. Thus, higher system reliability and availability are achieved through networking.

Information exchange over the network is another significant contribution available for the computer users. The networking capabilities have facilitated the evolutionary growth of information exchange among the academicians. Teleconferencing systems dealing with various research topics can act as a catalyst in reducing the cycle time for scientific results. This new form of communication medium can, in fact, go beyond the borders of the local campus through wide area networking. Focusing the attention of researchers throughout different universities on a certain research topic has a definite synergistic impact on scientific achievements.

The proliferating need to communicate among researchers can be suitably fulfilled through the networking opportunities such as

1. Exchange of messages,
2. Exchange of scientific papers and documents,
3. Transfer of data files reflecting experimentation results,
4. Transfer of source and binary code of application programs,
5. Interactive messaging and computer conferencing.

Remote login capability also enables users to access and use the remote facilities in an interactive or batch mode as it is desired.

This study is intended to illuminate the basic notions of networking, demonstrate the motivations in this direction, expose the related efforts and comment on the future trends. Section 2 presents a general picture of the distributed computer environment. Related queueing models are reviewed in Section 3. Local area network technology, wide area networking, and standardization efforts are elaborated in Sections 4, 5, and 6 respectively.

2. DISTRIBUTED COMPUTING ENVIRONMENT

As a consequence of the increase in the computing power of the personal systems and the decrease in their prices, it is ideal to distribute processing power to the place where it is needed. Distributing computing power does not mean to introduce unnecessary overheads. There is definitely economies of scale in keeping certain functions and expensive peripherals as centralized. Considering the secondary storage devices, for instance, the centralized large storage devices can provide service to a broader user community much more effectively than do the smaller ones connected to personal systems. Similarly, high quality and very fast printers should also remain centralized due to their purchase and maintenance costs.

Networking technology effectively connects users to several systems through their own personal workstation. Recently, many university campuses have been wired with a high capacity backbone network giving access to a large spectrum of computer resources on campus as well as national and international networks. A survey of 10 such universities and colleges are presented as case studies in a publication supported by EDUCOM Networking and Telecommunications Task Force (Arms (1988)).

The computing infrastructure of a distributed environment is based on the client-server model of Figure 1. A server is a centrally located machine to accomplish specific functions requested from other devices on the network. A client may use any one of the services provided. A client is not necessarily a personal computer. It can in fact be any I/O device on the network such as a laboratory instrument or a bar code reader. In general, server and client can both be programs. A server program most commonly runs on a dedicated computer. That is why the machine itself is treated as a server. A server can even utilize another service in which case it behaves as a temporary client.

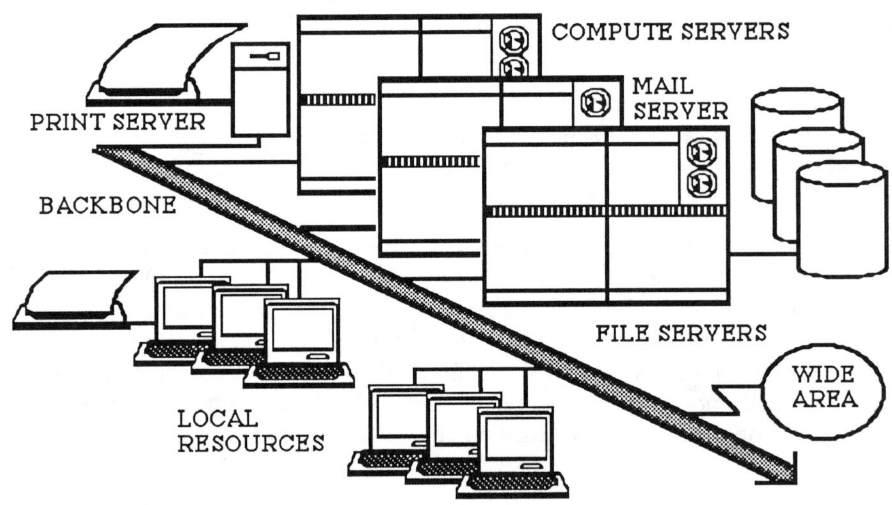

Figure 1. Client-Server Model

End users solve their computing problems, run any application, without having to know where the processing power comes from, what information is necessary, or what servers are involved in getting to the solution. The only computing environment that they need to be familiar with, is their own desktop workstation. They would not need to know where their applications software actually executes or where the information it generates really resides. Thus, a request from a user's device is satisfied transparently with processing power and data distributed throughout the network. Servers that are commonly implemented on most of the campus networks are as follows:

Print Server: They offer high speed and high quality printing services to clients on the network. Users access remote printers as if they are connected to the local workstations.

File Server: They provide transparent, remote file storage, access and sharing services to clients. Access to the user's personal files as well as public data files appears to be local to the user even though they are being delivered from the servers. Since all files are saved on a file server, it is not absolutely essential for the personal computers to have any local disks. This cost-effective implementation contributes advantages such as easier management, improved security and higher file sharing.

Mail Server: Electronic mail and messaging services allow users and application programs to exchange mail or messages in a network of non-homogeneous machines. These services are widely utilized by faculty and students on university campuses. This facility extends the faculty-student interaction outside the regular classroom and office hours. Assignments to students may be given and returned through this service. Coauthored papers with faculty members from other universities may be conveniently prepared using such a facility. Since the personal computers may be turned off when the mail arrives, there must be a mail server which is always active to handle mail. This server permits the users to read or send their mail as if the mail is held locally on the desktop.

Communication Server: Terminals may be linked to the network through a terminal server and may access any of the host computers on the network. Allowing users to access all the computer resources available on the network through the same terminal is much more cost effective and convenient than directly connecting the terminal to just one computer. Another communication server is a more complex linking device referred as gateway. This is a specialized hardware and software package to link between two dissimilar networks.

Computation Server: Referred also as "Compute Servers", these are powerful computers dedicated for computationally intensive applications. For example, a computer aided design application running on the user's own diskless workstation may benefit from such a server to quickly process the related data and the results may be locally displayed.

Database Server: This server is employed for storage and access of a database. In such an implementation, the application is split between the personal computer and the server. Local data handling of issuing requests for information and receiving that information is performed on the client machine whereas processing is carried on the server. For example, a typical database application in a campus environment is an on-line library catalog system. The system should allow the users to search by author, title, subject, and publication year. It should also facilitate to copy and paste citations and edit them into personal bibliographies. Public information services such as electronic bulletin board of scheduled academic events, jobs available for the new graduates as well as on-campus jobs for students, on-line information for the available computer services on campus should be easily accessible for the faculty, staff and students.

Name Server: User names and the names of the computer resources on the network should be kept in a continuously updated directory available on a centralized machine. This facility maps a user or a computer name to its correct network address. Managing these directories centrally is much more convenient and effective.

3. QUEUEING MODELING OF COMPUTER NETWORKS

Effective design and analysis of computer networks is a complex task. Theoretical results of queueing networks are helpful under certain simplifying assumptions. This section illustrates one of these design problems in a queueing context, and portrays the related optimization model. The reader is referred to Kleinrock (1976) for a more detailed exposition of the topic.

The communication network is physically composed of switching computers and communication channels which are depicted in Figure 2 as nodes and arcs respectively. Messages which are originating from a certain computer facility are first partitioned into packets. Each packet contains, besides its own content, the pertinent information such as source, destination, and priority class. It travels in a store-and-forward fashion on its route to the destination, hopping from one to the next switching point.

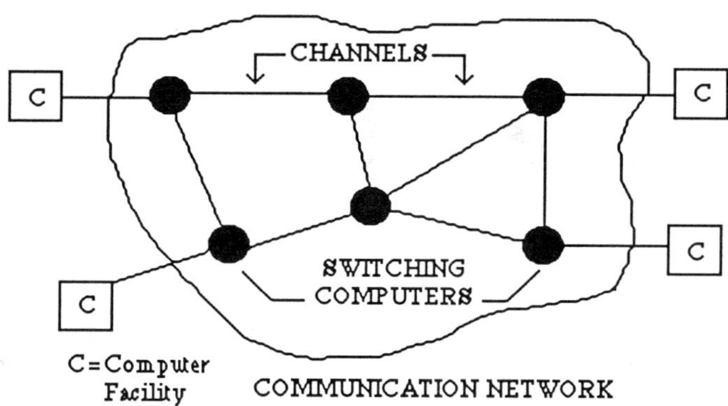

Figure 2. Structure of Computer Communication Network

The size of the problem of selecting the topological configuration grows rapidly as the number of nodes n increases. There are potentially $n(n-1)/2$ possible transmission lines and this leads to $2\,n(n-1)/2$ different configurations to be considered. One typical problem of capacity assignment for a given topology of an n-node, m-channel network under a specified traffic load may be posed as

Minimize　　Mean Message Delay Time
　　　　　　　with respect to the channel capacities
s.t.　　　　　　a total fixed budget.

One can similarly formulate the problem to minimize the total cost subject to a constraint of meeting certain performance measure of delay time. The channel capacities are the decision variables which normally take values in a discrete set. Network reliability which does not take part in this formulation is another important performance measure to be considered. There are quite a few other design questions of communication networks such as:

1. Selection of topological configuration,
2. Message routing procedure,
3. Flow control procedure,
4. Storage capacity allocation at each node,
5. Packet size selection,
6. Priority queueing discipline.

For the delay analysis of aforementioned capacity assignment problem, the underlying assumptions may be summarized as follows:

1. Packet arrival process to node i is Poisson with rate λ_i packets/sec,
2. Packet lengths are exponentially distributed with mean $1/\mu$ bits,
3. Message flow is according to a fixed routing procedure. That is, for any given origin-destination pair, there exists a unique path through the network,
4. All nodes have unlimited storage capacity,
5. Propagation delay which is the time required for the message bits to propagate down the length of the channel and the nodal processing times are negligible,
6. Packets are processed with the queueing discipline of first come, first served.

The illustrated problem where packets and channels are viewed as customers and servers respectfully, looks very similar to a Jackson open network with one major distinction. For a given message, the service times at different channels are not independent of each other. They are obviously dependent on the message length and the fixed channel capacity. Kleinrock makes the independence assumption for networks of moderate size and treats each channel as an isolated queueing station representable by an $M/M/1$ system.

For an n-node, m-channel communication network, define

$\lambda_i = $ Arrival rate for channel i in packets/sec,
$1/\mu = $ Expected packet size in bits/packet,
$C_i = $ Capacity of communication channel i in bits/packet,
$T_i = $ Expected message delay, i.e., service time, at channel i including the waiting time,
$\delta_{ij} = $ Traffic rate of messages originating from node i and destined to node j in packets/sec.

It is noted that the total external traffic of the communication network is

$$\delta = \sum_{i=1}^{n} \sum_{j=1}^{n} \delta_{ij}$$

and the expected total traffic within the network on all transmission lines is

$$\lambda = \sum_{i=1}^{m} \lambda_i$$

which may be much larger than δ. This is certainly true since a traffic load of 4 packets/sec from node A to D on the route A-B-C-D contributes this load on all of the three channels (AB,BC,CD) and is counted as 12 packets/sec in total. Noting that the service rate is μC_i packets/sec, the $M/M/1$ queueing system yields the result

$$T_i = \frac{1}{\mu C_i - \lambda_i} \quad \text{seconds.}$$

Proportion of traffic on channel i is

$$p_i = \frac{\lambda_i}{\lambda}$$

and mean delay per channel is

$$\sum_{i=1}^{m} p_i T_i \quad .$$

Mean number of hops per packet is

$$\bar{n} = \lambda/\delta.$$

Therefore, the expected message delay is

$$T = \bar{n} \sum_{i=1}^{m} p_i T_i \quad \text{seconds.}$$

If we denote d_i as the cost incurred for each unit of capacity built into the ith channel, the mathematical model of capacity assignment problem with linear cost structure is expressed as

Minimize $\quad T$

s.t. $\quad\quad\quad D = \sum_{i=1}^{m} d_i C_i.$

The Lagrangian function is formed as

$$G = T + \beta \left[\sum_{i=1}^{m} d_i C_i - D \right]$$

and the optimal solution is

$$C_i = \left(\frac{\lambda_i}{\mu}\right) + \left[\frac{\sqrt{\lambda_i d_i}}{\sum_{i=1}^{m} \sqrt{\lambda_i d_i}}\right]\left(\frac{E}{d_i}\right), \qquad i = 1, 2, \ldots, m$$

where

$$E = D - \sum_{i=1}^{m} \frac{\lambda_i d_i}{\mu}.$$

The traffic flow assignment problem and both problems combined are treated in Kleinrock (1976).

4. LOCAL AREA NETWORK TECHNOLOGY

Even though the early advances in networking were realized in long-distance or wide-area data communications, the local network technology has evolved rapidly in recent years. Local area networks (LANs) are communication networks which are required to interconnect a variety of devices within a small area. It is distinguished from other types of networks in that the communication is confined geographically to a small to moderate size area such as a single office building, a factory or a university campus. The distances covered are generally a few kilometers or less. A second significant characteristic is that a LAN is usually privately owned and used by a single organization.

Technological trends of local networks reflect high data rates and low error rates. Total bandwidths in the region of 1 to 10 Mbits/second are commonly available though recent technology is pushing the upper bound of this range to 100 Mbits/second. Even at such high speeds, error rate is consistently low as 10^{-8} to 10^{-11} bit error rate (BER). Another technological trend is the continuing decrease in cost and increase in capability of microprocessors utilized in network components.

The principal characteristics of LANs that determine its performance and cost are its topology, transmission medium, and medium access control protocol. Figure 3 portrays four different topologies as bus, ring, star, and tree. A bus is a long cable to which devices are connected. If you close the loop by joining the two ends of the bus, a ring topology is formed. In the star structure, each device is independently connected to the center. A tree can be viewed as an extended star structure, with the root acting as the central node, and its branches further have subbranches.

The traditional time sharing computer system with numerous terminal attachments has a star topology. For such a system, network reliability is absolutely dependent on the central computer. When it fails, no user on the network have access to any resources. It is not desirable in a local area network to totally rely on one central resource. Therefore, bus and ring topologies are more common. There is of course an advantage of star structure for troubleshooting in case of failure and ease of modification on the network if desired. Each node can be easily detached at the central site in order to isolate the problem without any need to physically trace the whole network. In practice, network designers take advantage of this convenience and form physical stars around flexible wiring boxes and attain logical configurations of buses or rings.

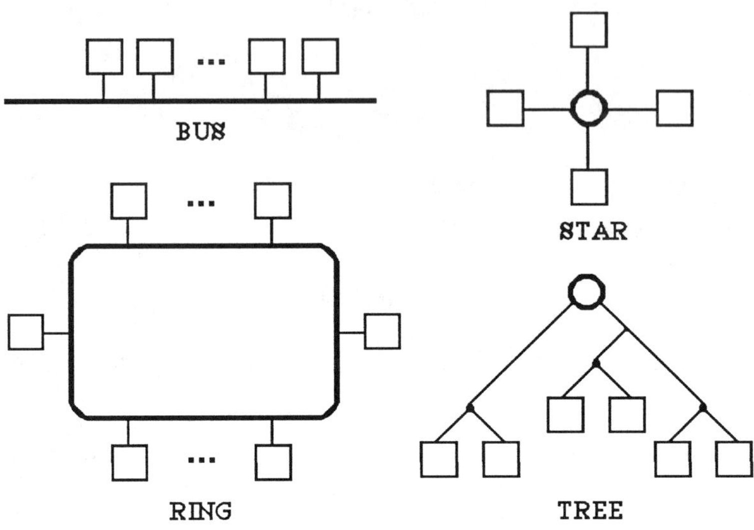

Figure 3. LAN Topologies

To select the most appropriate transmission medium for a LAN, one also has to consider, apart from the required performance properties related to speed and capacity, the total cost including the network interface equipment, installation labor and the cable material. Three different types of cable are generally used for LANs: twisted pair, coaxial, and fiber optic cables. For low speed transmission, twisted pair wiring is commonly used. Its weakness of interference and noise effects can be reduced by proper shielding. Data rate achievable over the twisted pair has increased dramatically in recent years. Coaxial cable is employed to achieve higher performance requirements. Higher data rates, longer physical distances and larger number of devices to be supported are the desirable characteristics of coaxial cables. This is the traditional medium for baseband ethernet. Fiber optic cables are very promising for the high speed connections as the backbone of the LAN installation. They carry signals as light pulses, rather than electricity. Very high data rates, like 80 to 100 Mbits/second, could be achieved. There are other advantages over the twisted pair and coaxial cable. Fiber is not sensitive to electrical noise and interference. It also has higher security features.

Access protocols to the transmission medium in LANS may be random or regulated. In random access, stations attempt to access at random; whereas the sequence and access times are governed deterministically by an algorithm for the regulated case.

For the bus topology, carrier sense multiple access with collision detection (CSMA /CD) is the most commonly used random access technique. A station wishing to transmit first listens to the medium to determine whether it is occupied with another transmission.

If there is no transmision in progress, the station attempts to transmit. Otherwise, it backs off for a certain period and retries afterwards. After transmitting, a station waits for a reasonable amount of time for an acknowledgement. Collisions only occur when more than one station begins transmitting within the period of propagation delay. Therefore, this protocol is quite suitable if packet transmission time is much longer than the propogation time. Token bus approach falls into the regulated access category. Stations on the bus form a logical ring among which a control packet known as the token circulates and this token governs the right of access. When a station receives a token, it is granted control of the medium for a specified time. When the station is finished with its work, or the due time has come, it passes the token on to the next station on the logical ring.

Similar to token bus, token ring is commonly practiced medium access control protocol for ring topology. The other well-known techniques for ring structure are register insertion and slotted ring which are both random access methods. For more information on LAN technology, the reader may refer to the tutorial text edited by Stallings (1988).

5. WIDE AREA NETWORKING FOR ACADEMICS

Knowledge acquisition, sharing, and dissemination are not restricted to a single campus. Researchers need to collaborate with their counterparts in similar other universities and research institutions. This section focuses on the networking efforts on national and international scales.

Sharing computing resources and communication over long distances on a national scale can be traced back to ARPAnet (Advanced Research Projects Agency Network) which was originally established in 1969 as a prototype packet switching network. Previously, circuit switching techniques was utilized in computer networks.

Before proceeding to give more information on ARPAnet, the distinction between these two switching techniques will be clarified. In circuit switching, a path is set up between the data terminal and destination. This path from source to destination is dedicated throughout the communication session until completion, analogous to telephone calls. The basic problem here is that all the branches along any given circuit are reserved for the whole call duration. Packet switching ,on the other hand, splits up long messages into a number of defined blocks or packets, and sends each individual packet as a separate entity. These packets can even follow different routes and get assembled at the destination node.

ARPAnet demonstrated the feasibility of a packet-switched network. It currently connects around 150 hosts in the United States. Most connections are over 56 Kbits/ second leased lines. The fundamental network protocols are the Transmission Control Protocol and the Internet Protocol, jointly abbreviated as TCP/IP. This protocol was adopted as the networking standard in 1982 by the Department of Defense in the United States.

In 1984, the National Science Foundation (NSF) launched a project to establish supercomputer centers and widen access to these NSF funded centers through a nation-wide computer communication system referred as NSFNET. The purpose of NSFNET is much broader than supercomputer access. The objectives can be summarized as:

1. Promoting access to NSF supercomputer centers,
2. Strengthening scientific collaboration among universities and research institutions,
3. Speed dissemination of research results,
4. Providing an experimental platform for research on network technology.

There is a three-level hierarchy in NSFNET: the backbone network, the mid-level networks, and the campus networks. The six supercomputer centers and the seven mid-level networks on the NSFNET backbone are listed in Table 1.

Table 1. Supercomputer Centers and Mid-Level Networks Attached to NSFNET Backbone in July 1989.

Supercomputer Centers:
Cornell National Supercomputer Center, Ithaca, NY
John von Neumann National Supercomputer Center, Princeton, NJ
National Center for Atmospheric Research, Boulder, CO
National Center for Supercomputer Applications, Champaign, IL
Pittsburgh Supercomputer Center, Pittsburgh,Pennsylvania
San Diego Supercomputing Center, San Diego, CA
Mid-Level Networks:
BARRNet, Bay Area Regional Research Network, Stanford University, Palo Alto, CA
Merit Computer Network, University of Michigan, Ann Arbor, MI
MIDnet (in six midwestern states), Univ. of Nebraska, Lincoln, NE
NorthWestNet (in northwestern states), University of Washington, Seattle, WA
Sesquinet, Rice University, Houston, TX
SURAnet, Southeastern Universities Research Association Network, University of Maryland, College Park, MD
Westnet, University of Utah, Salt Lake City, UT

NSFNET has grown rapidly in 1988-89. In July 1989, the number of connected campus networks has increased to 603 which was only 173 a year ago. The network is primarily based on packet switching technology and uses the TCP/IP family of protocols. The backbone of the network is substantially upgraded to 1.544 Mbits/second T1 circuits. This is equivalent to data transmission of approximately 50 pages of typed text per second. It also means about 30 fold increase over the pre-July 1988 backbone based on 56 Kbits/second links. The monthly traffic on this backbone has increased to 1.07 Billion packets in June 1989 according to statistics reported by Merit Inc., which is a membership consortium of eight universities in the State of Michigan in charge of the management and operation of the NSFNET backbone network.

BITNET (Because It's Time Network) was established between City University of New York (CUNY) and Yale University in May 1981. By the end of that year, six universities were connected. BITNET is open without restriction to any university. The philosophy is to create an easy to use, economical network for interuniversity communication. It may be used for any academic or administrative purpose. Therefore, the network has continued to expand rapidly. By fall 1987, the number of institutional members of BITNET had reached 360.

The network is based on leased lines linking the computers in different locations. Its technology is similar to IBM internal mail network (VNET). It operates with a protocol known as Remote Spooling Communications Subsystem (RSCS), and the line speeds are in general 9600 bits/second. The basic services of file transfer of all kinds, electronic mail, on-line messaging, remote submission of jobs to other computers are available. Actually, its speed is not very appropriate for any services that demand instantaneous response. It is based on message switching. That is, a file or message is sent to the next node over the link, stored there temporarily and forwarded over the next link. If the line is not available at any location, the file simply resides at that node until the line becomes available.

Clearly, the same needs for communication and cooperation between universities and research institutions existed in Europe. EARN, the European Academic and Research Network was started in late 1983. It is a general purpose computer network established with the clear aim of providing services to academic and research communities. It utilizes the same technology as BITNET. In three years, there were nineteen countries connected to EARN. About 600 computers located in approximately 300 institutions were linked. EARN is connected to BITNET with two links, forming a single logical network linking computers throughout Europe, North America, Japan, Israel, and Mexico. BITNET/EARN node table includes 2975 nodes in August 1989. This rapid growth leads us to think that eventually all scholars throughout the world will be joined for communication over an electronic network. Table 2 gives a picture of the network status for July 1989, indicating most of the countries connected and their corresponding node numbers. It is estimated that more than 70.000 academicians use this global network.

6. PROTOCOLS AND STANDARDIZATION EFFORTS

Protocols for computer network communication is a vital issue on university campuses. As it is illustrated in the previous sections, more computers are being interconnected hierarchically into one global network. Computer-based communication is gradually becoming an indispensable part of academic life. However, for a heterogeneous computer environment, incompatibilities and the lack of easy interworking is a big hindrance to provide harmonized data communication services to the end-users.

Protocols consist of all the rules that govern communication between computing systems. Standards are absolutely essential to achieve uniformity and encourage interoperability. If standards are not established, a substantial cost would be incurred for software development to convert from one computer system to another among protocols. The requirement for all systems to communicate with one another can be achieved by having a common interface to which all sets of protocols can map. Since it is hard to guarantee commonality of protocols across all systems, it is useful to have a reference model to which all systems can be converted.

Table 2. Most of the Countries Connected to BITNET/EARN and Their Node Numbers (July 1989)

Argentina	3	Austria	15
Belgium	31	Brazil	10
Canada	188	Chile	10
Denmark	18	Egypt	1
Finland	26	France	141
Germany	231	Great Britain	3
Greece	12	Iceland	1
Ireland	6	Israel	55
Italy	112	Japan	65
Korea	6	Luxemburg	1
Mexico	9	Netherlands	75
Norway	5	Portugal	2
S. Arabia	2	Singapore	6
Spain	25	Sweden	23
Switzerland	48	Taiwan	11
Turkey	13	USA	1794
Yugoslavia	1		

Open System Interconnection (OSI) is an international standards activity (Aschenbrenner (1986)). OSI was initiated in March 1977 by ISO, the International Organization for Standardization, and interest and participation in it have continued to grow. It primarily defines formats and protocols to interconnect systems that have different architectures provided by different suppliers. The principal objective of OSI is interconnection of systems of different vendors, and coordination of standardization activities in telecommunications and information systems.

The basis of OSI standardization is a seven layer reference model for the coordination of standards development. The top layer, which is called the application layer, provides the interface for performing real functions such as file transfer, electronic mail, or remote login. The bottom layer is referred as the physical layer, and it provides the mechanical and electrical interface for sending and receiving bit streams.

One aspect of the Open Systems Interconnection reference model is that it allows maximum interoperability among different network architectures. This is achieved not by having a common architecture, but by ensuring that gateways can be built to the standard reference architecture. Interworking of computer networks is of prime importance for several international organizations. RARE is the association of European research networks and their users. It was formally established in Amsterdam on June 13, 1986 with 14 national members and 3 international members. Its aim is to foster cooperation between both national and international network organizations in order to promote interworking and interoperability. The ultimate goal is to develop a harmonized international communications infrastructure to enable researchers to communicate and access computer resources throughout Europe and in other continents.

The COSINE (Cooperation for Open System Interconnection Networking in Europe) Project was adopted as a part of the Eureka programme in November 1985. The project aims to provide the European research community an open networking infrastructure. There are currently 19 participating countries in this project. The Cosine community is estimated to involve over 2500 institutions with a target user population of one million researchers including 50% support staff. The services will be based upon international ISO/ OSI standards and will use the common carrier services for basic conveyance of wide area traffic. RARE undertook technical work for the COSINE Specification Phase under contract to the COSINE Policy Group. The conclusion derived from this definition phase is that realization of a pan-European networking infrastructure based on the original proposal "The technical Specification of the COSINE Infrastructure" can be achieved (RARE (1988, 1989)).

Several other key activities have been launched in the past few years to address the need to improve software development environment in terms of applications portability. The Institute of Electrical and Electronics Engineers (IEEE) which has a long history of sponsoring standards efforts developedan open operating system standard called POSIX (Portable Operating System for Computer Environment).

Founded in Europe in 1984, X/Open is an independent group of international computer systems vendors who are investing significant resources in the development of an open, multi-vendor, common applications environment based on de facto and official international standards. X/Open describes a Common Applications Environment (CAE) with all the technical specifications related to operating system services, programming languages, data management, networking, window management, and security.

The Open Software Foundation is an international organization dedicated to the development and delivery of an open, portable software environment to which vendors and users have equal input and access. Open Software Foundation is established on May 17, 1988 originally sponsored by eight computer companies and its current membership is comprised of nearly 100 organizations, including academic institutions, user organizations and research foundations, and private companies from a broad range of industries. OSF's goal is to make it easier for users to mix and match software and hardware from different suppliers, creating a vendor neutral software architecture in which computer systems from many suppliers work together in a virtually seamless environment

7. CONCLUDING REMARKS

Networking environment creates a global computer system where the whole is greater than the sum of the parts. Users access all the resources contained in a computer network from their own desktop. This is achieved transparently providing users a single-system image of the storage, applications and processing power of the entire cluster from their own personal system. The net result of this seamless user interface is improved productivity through reduced learning time and a significant decrease in user errors allowing the application to be run wherever it is most appropriate; smaller tasks on workstations and larger problems on more powerful systems.

Having a cohesive software architecture across all levels of computer platforms and with strong network underpinnings is a sound and sensible strategy. There is a growing interest in open systems. Computer users want to preserve their applications and their data as their computer networks expand with additional resources from different manufacturers. The concepts most important to open systems are portability, scalability, and interoperability.

Portability is the ability to use the application software on computers from different vendors, without rewriting the program's code. Scalability refers to a user's ability to move applications and data among larger and smaller computer systems to meet changing needs. Users can move applications among stand-alone personal systems, multiuser mini systems, and powerful mainframes without changing their operating system, the application, or their user interface. This idea of having a broad range of processing power under the same operating system is very appealing. Interoperability refers to the ability to run applications programs on networks built up of different machines by different vendors. The key benefit of open systems is that applications and user interfaces remain the same, even though the computer networks combine different kinds of hardware in a multivendor heterogeneous environment.

Future academic computer environment will connect more scholars and enable them to utilize computer applications in an easy and consistent way. This will bring a significant impact on the research and teaching productivity. Hence, academicians will be able to accomplish more work in less time.

8. REFERENCES

1. Arms,C. 1988. Campus Networking Strategies. In *EDUCOM Strategies Series on Information Technology* , C. Arms(Ed.), Digital Press, Bedford, Massachusettes.

2. Aschenbrenner, J.R. 1986. Open System Interconnection. *IBM Systems Journal* **25**, 369-379.

3. Hanss, T. 1989. NSFNET, Seminar Notes of Distributed Computing. IBM Europe Institute .

4. Kaylan, A.R. 1990. Queueing Networks: A Survey of Analytical Results. In *Queueing Theory and Applications* , S.Özekici (Ed.), Hemisphere, New York.

5. Kleinrock, L. 1976. *Queueing Systems: Computer Applications.* Vol.2. Wiley, New York.

6. Lord, D. June 1985. Worldwide Networking for Academics. *Data Processing* **27**, 27-31.

7. RARE. 1988. COSINE: Cooperation for Open Systems Interconnection Networking in Europe. RARE, Amsterdam.

8. RARE. 1989. RARE Annual Report 1988. RARE, Amsterdam.

9. Rounds, T.R. 1989. CNSF Experiences with Workstation-Supercomputer Distributed Visualization, Seminar notes of Distributed Computing. IBM Europe Institute.

10. Stallings, W. (Ed.) 1988. *Local Network Technology.* Computer Society Press, Washington,D.C..

11. Stallings, W. 1989. The Glue for Internetworking. *Byte* **14**, 221-224.

12. Tanenbaum, A. S 1981. *Computer Networks.* Prentice Hall, Englewood Cliffs, New Jersey.

MANAGEMENT AND OPTIMIZATION
OF
QUEUEING SYSTEMS

Stanley R. Pliska
Department of Information and Decision Sciences
University of Illinois at Chicago
Chicago, Ill. 60680, USA

1. INTRODUCTION

The purpose of this study is to provide a survey of operations research results on the optimal management of queueing systems. The emphasis is on the results themselves and how these results may be useful for practical problems , rather than on the theoretical derivation of these results. Nevertheless, attention will be given to the fundamental principles that are used in this subject area.

Roughly speaking, the subject area is divided into two parts: static, optimal design issues versus dynamic, optimal control issues. In the first part various system configurations are considered and the analysis of the resulting equilibrium, steady-state behavior allows one to determine the best system for optimizing some long-run criterion such as average annual cost or profit. The term "static" is used because once the system configuration is chosen, then it does not change over time. For example, a system may be designed to have three parallel servers, and then once this system is developed, it continues with three servers indefinitely.

The second part of the subject area, the study of dynamic, optimal control issues, pertains to situations where characteristics of the system are allowed to change over time. The term "dynamic" refers to the fact that the optimal choice of some characteristic (such as the number of servers) may be allowed to vary according to some state of the system (such as the number of waiting customers) which, in turn, would normally vary over time. Dynamic programming methods, rather than steady-state distributions, tend to be used to determine the optimal decisions for these dynamic situations.

Another way to distinguish between the two kinds of management problems for queueing systems is to think of the computer analogy of the distinction between hardware and software. The static design issues often pertain to queueing system characteristics that, like computer hardware, are normally not changed once the investment has been made. On the other hand, dynamic control issues pertain to characteristics that, like computer software, are easier to change as time goes on; moreover, one may wish to

168

use the best software for the computer equipment available.

Having said all this, it should be pointed out that some practical management problems may involve both static and dynamic issues, thereby making it difficult to classify queueing theory results that are needed. Nevertheless, the reader should try to keep this distinction in mind in order to facilitate the learning of this subject.

It should also be mentioned that the literature on dynamic control issues is much larger than the literature on static design issues. This is because the methodology required for determining optimal designs is relatively straightforward: one simply applies the already-established formulas for the steady-state distributions. Such analyses are often too simple to deserve publication as separate research papers; the basic principles and results are of a general nature and can be presented in a relatively small number of papers. On the other hand, to determine optimal decisions for dynamic situations it is usually necessary to solve complicated dynamic programming models that vary from one specific problem to another. Each solution of a specific dynamic control problem is unique and usually is sufficiently complicated to deserve a separate paper.

While the majority of relevant references are of a specific nature, a few of general interest should be mentioned. Crabill, Gross, and Magazine (1977) compiled a bibliography of research on the optimal design and control of queues. Two survey papers on dynamic control issues were presented at the same conference in 1973: Prabhu and Stidham (1974) surveys the use of the semimarkov decision theory methodology for queueing control, and Sobel (1974) focuses on the structure of optimal policies for the service rate. These three sources list about 200 different references. No special attempt will be made to repeat those references here, but an effort will be made to identify more recent references. One is the book by Heyman and Sobel (1982), which has a chapter called "System Properties" that is relevant to queueing system management.

Before diving into the research literature we will look at a case study. This is a model of telephone operator staffing that has actually been used in the telephone industry to reduce the cost of meeting service criteria, to help with planning, and to help explain to regulatory agencies changes in the service measurement criteria. Hopefully, the understanding of this example will facilitate the subsequent learning of static design and dynamic control methodologies.

The balance of this paper is devoted to the basic principles of queueing system management and is divided into three sections. Two sections deal with static design and dynamic control issues, respectively, both for conventional queueing systems where the customers depart the system as soon as their initial service is completed. In contrast, the final section deals with the management of queueing networks.

2. CASE STUDY:TELEPHONE OPERATOR STAFFING

This case, taken from "A Queueing Model For Telephone Operator Staffing" by Sze (1984), is an example of how to use an equilibrium model for optimal design. In particular, the problem is to minimize the number of telephone operators (that is, the number of servers) subject to two constraints: (1) the average customer waiting time W is less than a specified number (which is usually between 1 and 7 seconds, depending on the situation) and (2) W is less than 10 seconds when the Poisson arrival rate λ_0 is increased by 5%. The approach is to simply use a standard $M/G/s$ model to calculate W for these two values of λ_0 and a particular value for the number of operators s. If a constraint is violated, then the calculations are repeated for a larger value of s. If both constraints are slack by a sufficiently wide margin, then the calculations are repeated for a smaller value of s. By proceeding through a logical sequence of values of s, this procedure converges to the optimal value of s.

Six special features make this case study especially interesting:

1. While the arrivals can be modeled as a Poisson process, the arrival rate varies considerably over time.
2. Often 100 or more operators are required, so one needs to compute $M/G/s$ formulas for larger values of s than is common in the literature.
3. The constraints are such that the results for neither heavy traffic nor light traffic systems can be applied (so no shortcuts).
4. The service time distribution (denoted B) is often such that Erlang approximations are inappropriate (no shortcuts here either).
5. Customers who are delayed may abandon their calls ($A(t)$, the distribution function of a customer's maximum waiting time T, is known).
6. Customers who abandon their calls may reattempt them (this occurs with probability p, which is known).

To handle the first feature, Sze (1974) assumes that the arrival rate is constant over 30 minute intervals and that W for each such interval can be approximated by the W for the corresponding equilibrium, steady-state situation. The difficulty with regard to the remaining five features is the computation of W for particular values of the data λ_0, A, B, p, and s.

The flow of the customers is as shown in Figure 1, where the arcs are labeled with the flow rates and where the quantities $\lambda \equiv \lambda_0 + \lambda pq$ and $q \equiv P(\text{customer abandons call})$ have been introduced.

Let the random variable Z be the service load or waiting time in equilibrium or steady-state, so $W = E[Z]$. Let F denote the distribution function for Z, and note that

$$q = P(T < Z) = \int_0^\infty A(t)\, dF(t).$$

This means that the unknown quantity q depends on the unknown distribution F. Meanwhile, looking at Figure 1, one sees that if one knows the effective arrival rate $\lambda(1 - q)$, then one can use standard queueing results to compute the distribution F. Hence, in effect, we have three equations and three unknowns: λ, q, and F. But Sze explains how to use a recursive procedure that converges to the solution of these three equations. Finally, knowing F one computes $W = E[Z]$, the measure of performance for the case being evaluated.

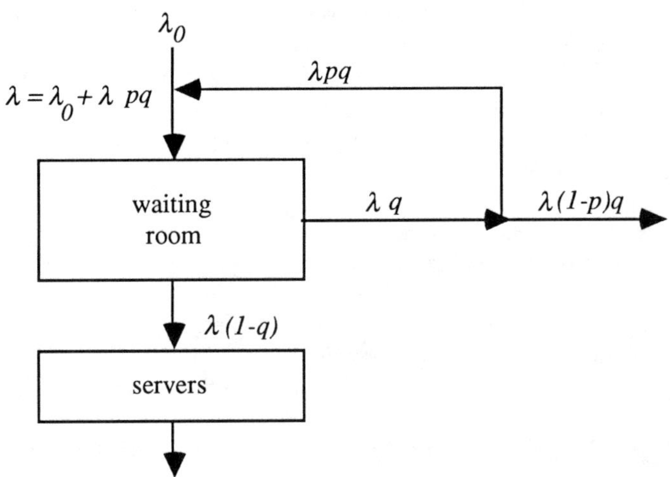

Figure 1. The Flow of Customers in the Case Study

Sze (1984) validated his approach for computing W with field data from three cities. He looked at several hundred half-hourly intervals for each city, and for each interval he compared the observed and computed values of W. He then did a regression analysis for each city, hoping the regression line would have slope of 1 and intercept of 0. The actual slopes (respectively, intercepts) he got were 1.139, 0.900, and 0.728 (resp. 0.037, 0.04, 0.112). The t statistics were marginally significant.

3. STATIC OPTIMAL DESIGN PROBLEMS

The telephone operator staffing case that was studied in the preceeding section illustrates a typical static, optimal design problem. In general, the approach is to formulate a fairly standard optimization problem, with the objective function and the constraints being simple expressions that involve steady-state, equilibrium quantities such as the average waiting time W. Parameters such as the arrival rate λ, the service rate μ, and the number of servers s are the decision variables. Since steady-state, equilibrium quantities are involved, this approach is most accurate for situations where the system's characteristics are constant over time. However, as we saw for the telephone operator staffing problem, this approach can also be used for situations where the system's characteristics vary over time: one assumes that during short periods the system's behavior can be approximated by steady-state expressions. For example, the average number of customers waiting in line during the opening hour at a bank can be approximated by the standard, steady-state expression, even though the customer arrival rate later in the day will be much different.

For static, optimal design problems the objective function is often cost or profit.

One example is to choose the number of servers s so as to minimize the cost $c_w L + c_s s$, where c_w is the waiting cost per customer per unit time, c_s is the operating cost per server per unit time, and the steady-state quantity L is the average number of customers in the system. Note that L is actually a function of s.

For example, suppose for an $M/M/s$ system the problem is to choose a value of s equal to 1, 2, or 3. Moreover, suppose $c_w = 1$, $c_s = 2$, and $\rho = \lambda/\mu = 0.8$. Using the standard equation for L one computes $L = 4$, 1.16, and 0.8 for $s = 1$, 2, and 3, respectively. Hence the objective function values are 10, 5.16, and 6.8, respectively, in which case $s = 2$ is the optimal choice.

A slightly different situation is to choose the service rate μ for a single server system so as to minimize the cost $c_w L + c(\mu)$, where $c(\mu)$ is the operating cost per unit time as a function of μ. Now L depends on μ. With problems like this, one might assume that the decision parameter must take one of a finite number of specified values, in which case one would simply compare a finite number of objective function values to determine the best design. Alternatively, one might assume that the decision parameter can take all the values in some interval.

For example, suppose one can choose any $\mu > \lambda$ for an $M/M/1$ system with $c(\mu) = c\mu$. Since $L = \lambda/(\mu - \lambda)$, the objective function is $c_w \lambda/(\mu - \lambda) + c\mu$. Differentiating this with respect to μ shows this is minimized by the value

$$\mu = \lambda + \sqrt{c_w \lambda/c}.$$

A rather different kind of design problem is to organize the customers into priority categories so as to reduce the average number of customers in the system. This is done by giving higher priority to customers who can be recognized in advance as having shorter processing or service times. Just as with the shortest processing time discipline for job shop scheduling (see the book by Conway, Maxwell, and Miller), by promptly serving those customers who can be quickly dispatched, the average waiting time for all the customers can often be reduced. Of course, this is a deviation from the first-come-first-served discipline, and the average waiting time for the more time-consuming customers may be greater than under the conventional FCFS discipline.

In order to implement this shortest processing time idea, it is, of course, necessary to be able to estimate a customer's service time in advance. But this requirement is frequently satisfied. For example, U.S. supermarkets often have a check-out line reserved for customers purchasing fewer than 8, say, grocery items. Compared to the conventional situation where all the check-out lines are the same, the express-line customers have much lower waiting times, thereby offsetting the slightly increased waiting times for the customer with more items to purchase. Other situations where the shortest processing time concept can be employed include airline check-in (express line if you already have your ticket), and hospital emergency rooms (for minor emergencies, give higher priority to those that can be treated more quickly).

Perhaps the best known reference on priority queues is the book by Jaiswal. Unfortunately, formulas for the average waiting time for each priority class of customers are only known for a small number of simple situations. It is therefore often necessary to rely on computer simulation. In particular, one would first specify a finite number of priority organization schemes, then estimate the average waiting times for each scheme with computer simulation, and finally select the best scheme.

Another kind of design problem concerns the trade-off for multi-server systems of having a single waiting line versus individual waiting lines for each server. The choice is often available for banks and government offices. Having individual lines often requires less space and is a system that is easier to implement, but the average waiting times may be greater. Furthermore, with a single line the customers are more satisfied knowing the FCFS discipline is adhered to. In order to help analyze this choice one would simply compare the single-line model, which has arrival rate λ and s servers, with a model for just one line in the individual-line case, a model that has arrival rate λ/s and a single server. Of course, for the individual-line case this does not model how customers might switch lines as they see other lines becoming shorter than theirs, so the computed waiting times will be biased too high.

We finish our look at static design problems with a discussion of circumstances when it is better to have only one server than a larger number. This is based on the work by Stidham (1970). The problem is to choose for a $GI/M/s$ queueing system the number s as well as the service rate μ so as to minimize the average cost per unit time, namely, $Ks\mu + CL$. Here L is the average number of customers in the system and K and C are positive constants, so $Ks\mu$ is the cost rate associated with system capacity $s\mu$, while CL is the waiting cost rate.

Consider an arbitrary pair $(\mu,\ s)$, and compare this choice with the single-server system having service rate $s\mu$. The system capacity cost rate for these two choices is the same, namely, $Ks\mu$. But it can be shown that the average number of customers in the single-server system is no greater than the average number in the multi-server system. This is true because the total service capacity $s\mu$ in the single-server system is always utilized whenever customers are present. However, in a multi-server system with the same capacity, the capacity will not be utilized whenever the number of customers is less than the number of servers. In particular, the multi-server system with $n < s$ customers will have a lower output rate than the single-server system with the same number of customers, and this leads to a lower average number of customers in the single-server case.

Although Stidham's idea about the advantage of single-server systems is interesting, it is probably difficult to implement in practice. For example, if a bank requires five normal tellers, then it may be impossible to find a "super-teller" whose service rate is five times faster than normal. However, when you look at Stidham's idea in a more general way, then you have an important principle: when adding resources to your queueing system in order to improve performance, it is often better to speed up your current servers than to add additional servers.

4. DYNAMIC OPTIMAL CONTROL

As mentioned in the introduction, it is often appropriate to change a queueing system's characteristics in a dynamic manner, depending on how other characteristics might be varying with time. For example, it may be appropriate to add additional servers or to speed up service if the number of waiting customers becomes sufficiently large, and then resume normal operations once the number waiting has diminished. This section addresses these kinds of problems.

It is convenient to divide dynamic, optimal control problems into two categories: control of the arrival process and control of the service process. For arrival process control there are three main possibilities. One is to simply deny admission to entering customers under certain circumstances. In other words, this is a binary situation where sometimes you "shut the door". Another possibility is to physically control the arrival rate, such as by rejecting every second arrival when the system becomes crowded. A third possibility is to impose pricing policies or tolls in such a way as to suitably affect the arrival rate, even though, in contrast to the second possibility, each customer makes the decision of whether to enter the system.

The other kind of dynamic control problem involves controlling the service process. Here there are also three main possibilities. The first is to add servers when the system becomes crowded. The second is to keep the same number of servers but vary the service rate for each server. The third subcategory involves batch service systems, namely, deciding how long to wait prior to serving (or dispatching) a group of customers.

One thing most all these control problems have in common is that the decision variable will be expressed as a function of the number of customers in the system. Since the number of customers (the "state" of the system) will vary in a dynamic manner, so will the value of the decision variable. The hard part is finding the best control function, that is, finding the best rule that specifies the values of the decision variable as a function of the number of customers.

There are two main approaches for finding the best control function. The first, called the "direct" approach, is to specify some control functions and then analyze each with standard queueing theory methods. For example, suppose the control function says to admit all arriving customers if there are fewer than k customers in the system, and otherwise randomly turn away 50% of the arriving customers, where k is any integer. Then with Poisson arrivals, exponential service, and a single server, one could analyze a birth-and-death model to determine various system characteristics as a function of k, thereby leading to the determination of the control function having the best k. This will be illustrated below.

The second approach is to assume little about the form of the optimal control function but to use dynamic programming methods to compute the optimal control function. The advantage of this "dynamic programming" approach is that a larger class of control functions will automatically be considered, and so the computed control function will often be superior, and never inferior, to that derived with the direct approach. For example, in the case of the arrival process control problem stated above, dynamic programming may reveal that it is better to let the turn-away percentage for some states be other than 0% or 50%. The disadvantage of dynamic programming, however, is that the computations and analysis are often more complicated.

The balance of this section is devoted to four examples. First, we see how the direct approach can be used to solve the arrival control problem that is stated just above. Second, we introduce dynamic programming by solving a simple discrete time, batch service problem. Then we look at how to use dynamic programming in a more advanced way to decide when a server should be operating an $M/G/1$ system. Finally, we look at optimal policies for batch service queueing systems.

For the arrival control problem, consider an $M/M/1$ system with arrival rate λ and service rate μ, and suppose arriving customers are randomly turned away with probability 50% whenever k or more customers are already in the system. Then while k or more customers are in the system, the process describing the actual additions to the system is Poisson with parameter $\lambda/2$. Hence this system can be modeled as a birth-and-death process with arrival rate λ for each state $n < k$, arrival rate $\lambda/2$ for each state $n \geq k$, and service rate μ for each state $n \geq 1$.

The steady-state, equilibrium distribution will exist if $\rho \equiv \lambda/2\mu < 1$. Assuming this to be the case, one can use the standard approach to solve for the steady-state probabilities P_n, $n \geq 0$. First, one derives an expression for P_n in terms of P_0, namely

$$P_n = \begin{cases} 2^n \rho^n P_0, & n = 0, 1, .., k \\ \rho^{n-k} P_k = 2^k \rho^k \rho^{n-k} P_0, & n = k, k{+}1, .. \end{cases}$$

Then since $1 = P_0 + P_1 + ...,$ we have

$$\begin{aligned} 1 &= P_0 \left[1 + 2\rho + ... + 2^{k-1}\rho^{k-1} + 2^k \rho^k \{1 + \rho + \rho^2 + ...\}\right] \\ &= P_0 \left[1 + 2\rho + ... + 2^{k-1}\rho^{k-1} + 2^k \rho^k/(1 - \rho)\right]. \end{aligned}$$

This means

$$P_0 = \left[1 + 2\rho + ... + 2^{k-1}\rho^{k-1} + 2^k \rho^k/(1 - \rho)\right]^{-1},$$

and substituting this in the above expression for P_n completes the calculation of the steady-state distribution. /// Now suppose the queueing system collects a reward r each time a customer is served, but the waiting cost c is incurred per unit time for each customer in the system. For a particular value of k we are interested in computing $f(k)$, the profit per unit time. This equals

$$f(k) = r\mu^* - cL,$$

where L is the average number of customers in the system and μ^* is the steady state processing rate. Both L and μ^* can be computed from the steady state probabilities. In particular,

$$L = \sum_{n=0}^{\infty} nP_n = P_0 \left[\sum_{n=0}^{k-1} n2^n \rho^n + 2^k \sum_{n=k}^{\infty} n\rho^n \right]$$

$$= P_0 \left[\sum_{n=0}^{k-1} n2^n \rho^n + 2^k \rho(1-\rho)^{-2} - 2^k \sum_{n=0}^{k-1} n\rho^n \right],$$

an expression that is not pretty but is straight-forward to compute. The processing rate μ^* is the same as the rate at which customers depart the system, that is, $\mu^* = \mu(1 - P_0)$. For example, with $k = 0$ one has $P_0 = 1 - \rho$, $L = \rho/(1-\rho)$, and $\mu^* = \lambda/2$, whereas with $k = 1$ one has $P_0 = (1-\rho)/(1+\rho)$, $L = 2\rho(1-\rho)^{-1}(1+\rho)^{-1}$, and $\mu^* = \lambda/(1+\rho)$.

To complete the solution of this problem one needs to find the value of k that maximizes the profit rate f. It is not clear whether f is concave with respect to k, so one may need to exhaustively search among many values of k in order to find the best one. Note that if $\rho > 1/2$, then $L \to \infty$ as $k \to \infty$, so in this case it would be a mistake to choose a large value of k. But if $\rho < 1/2$, then L is bounded above by $2\rho/(1 - 2\rho)$, and it is possible that it would be best to choose k very large.

The dynamic programming approach for solving dynamic control problems will be introduced by considering a simple discrete time, batch service system. Suppose A customers arrive each period, with $P(A = 0) = P(A = 1) = P(A = 2) = 1/3$ and with the arrivals being independent in different periods. After any such arrivals, the decision maker must choose one of two actions: either process all the customers at a cost of \$3 or process none and incur the holding cost of \$1 per customer. This cycle is repeated every period, but in the final period every customer present must be processed. The problem is to choose the optimal action, a choice that will depend on the number of customers present as well as the number of periods to go.

Dynamic programming involves the recursive computation of the optimal value function. In this case, this is $V(s, n)$, the expected cost over the n remaining periods if s customers are now present and optimal choices are made in the remaining periods. Since

$$V(s, 0) = \begin{cases} 0, & s = 0 \text{ (i.e., no processing necessary)} \\ 3, & s \neq 0 \text{ (i.e., process all customers)} \end{cases}$$

the first step is to compute $V(s, 1)$ for all s in terms of $V(s, 0)$. Since only one choice remains to be made, namely, the action in the next to last period, $V(s, 1)$ will equal the minimum of two expressions, each corresponding to the expected remaining cost under the associated action. For example,

$$V(1, 1) = min\{1 + \frac{1}{3}V(1, 0) + \frac{1}{3}V(2, 0) + \frac{1}{3}V(3, 0),$$

$$3 + \tfrac{1}{3}V(0, 0) + \tfrac{1}{3}V(1, 0) + \tfrac{1}{3}V(2, 0)\}$$

$$= min\{1 + \frac{1}{3}(3) + \frac{1}{3}(3) + \frac{1}{3}(3), \ 3 + \frac{1}{3}(0) + \frac{1}{3}(3) + \frac{1}{3}(3)\}$$

$$= min\{4, 5\} \ = \ 4.$$

Since the minimum 4 corresponds to the action "hold", we know this is the optimal action with $s = 1$ and one period to go. Similarly,

$$V(2, 1) \ = \ min\{2 + \frac{1}{3}V(2, 0) + \frac{1}{3}V(3, 0) + \frac{1}{3}V(40, 0), \ 5\}$$

$$= \ min\{5, 5\} \ = \ 5,$$

so there is a tie between the two actions as far as which one is optimal. For all $s \geq 3$ we have

$$V(s, 1) \ = \ min\{s + 3, 5\} \ = \ 5,$$

so it is optimal to process. Finally, if $s = 0$ then "process" is clearly not optimal, so

$$V(0, 1) \ = \ \frac{1}{3}V(0, 0) + \frac{1}{3}V(1, 0) + \frac{1}{3}V(2, 0) \ = \ 2.$$

To summarize up to this point, we have computed the optimal value function $V(s, 1)$ and the optimal actions with one period to go. The second step is to compute $V(s, 2)$ and determine the optimal actions with two periods to go. Here the dynamic programming idea is employed: $V(s, 2)$ equals the expected cost if the optimal action is used with two periods to go and then optimal actions are used the rest of the way. Thus $V(s, 2)$ equals the minimum of two expressions, each corresponding to one of the possible actions with two periods to go. Each expression involves two terms. The first is the cost associated with the chosen action and current period. The second term is the expected cost over the remaining periods; since optimal actions are used the rest of the way, this latter term can be expressed in terms of $V(s, 1)$, which was just computed. For example,

$$V(1, 2) \ = \ min\{1 + \frac{1}{3}V(1, 1) + \frac{1}{3}V(2, 1) + \frac{1}{3}V(3, 1),$$

$$3 + \tfrac{1}{3}V(0, 1) + \tfrac{1}{3}V(1, 1) + \tfrac{1}{3}V(2, 1)\}$$

$$= \ min\{1 + \frac{1}{3}(4) + \frac{1}{3}(5) + \frac{1}{3}(5), \ 3 + \frac{1}{3}(2) + \frac{1}{3}(4) + \frac{1}{3}(5)\}$$

$$= \ min\{\frac{17}{3}, \frac{20}{3}\} \ = \ \frac{17}{3},$$

so "hold" is the optimal action with two periods to go. With $s \geq 2$ we have

$$V(s, 2) \ = \ min\{s + \frac{1}{3}V(s, 1) + \frac{1}{3}V(s + 1, 1) + \frac{1}{3}V(s + 2, 1), \ \frac{20}{3}\}$$

$$= \ min\{s + 5, \ \frac{20}{3}\} \ = \ \frac{20}{3},$$

so it is optimal to process. Finally, with $s = 0$ it is clearly not optimal to process, so

$$V(0, 2) = \frac{1}{3}V(0, 1) + \frac{1}{3}V(1, 1) + \frac{1}{3}V(2, 1) = \frac{11}{3}.$$

Hence with 2 periods to go it is optimal to process right away if and only if two or more customers are present.

Having computed $V(s, 2)$ we are now ready for the third step, the computation of $V(s, 3)$. But the pattern is the same, so the details will be omitted. Suffice it to say that, in general, having computed $V(s, n)$ one next computes $V(s, n + 1)$ by using the dynamic programming principle described above. Hence the problem is solved in a recursive manner.

In order to study some more interesting dynamic optimal control applications to queueing problems it is necessary to introduce some additional dynamic programming concepts. If either the arrival process is Poisson or the service times are exponential (or both), then the dynamic programming approach involves what is called a *semimarkov decision process*. In particular, the queueing system control problem is formulated as a semimarkov decision model, and then the semimarkov decision theory is used to determine the optimal control function. Stidham and Prabhu (1974) wrote a survey paper that describes this approach. The second volume by Heyman and Sobel is a good reference for semimarkov decision theory, and it also is the reference for the service process control example that will soon be presented.

A semimarkov decision process is like a semimarkov process except that each time the process enters a new state, the decision maker chooses an action which affects the subsequent motion of the process and generates a reward. In particular, let S be the state space and let A be the set of admissible actions. Three data elements are needed:

1. A law of motion probability distribution P,
2. A sojourn time distribution function F,
3. A cost function c.

To see how this semimarkov decision process works, consider how it would be simulated:

1. The process enters state $s \in S$,
2. Action $a \in A$ is immediately selected,
3. Cost $c(s, a)$ is incurred,
4. The next state s' is generated according to the conditional probability measure $P(.|s, a)$,
5. The sojourn time τ in state s is generated according to the conditional probability distribution function $F(.|s, s', a)$,
6. After τ time units the process jumps to state s' and this cycle is repeated.

Many queueing control problems can be formulated in this way. One seeks a *policy* or rule that governs how to choose the actions. A policy is said to be *stationary* if this rule depends only on the state, that is, the choice of action might depend on the state that was entered but not on the time or any other aspect of the past history.

Each policy is evaluated with a measure of performance that is a function of the costs. Let π denote an arbitrary policy, and let the stochastic process Z_t be the corresponding cumulative costs during $[0, t]$. One commonly used measure of performance is the expected discounted cost over an infinite planning horizon:

$$v(\pi, s) = E\left[\int_0^\infty e^{-\alpha t} dZ_t \,|\text{initial state is } s\right];$$

here $\alpha > 0$ is the discount rate. Another commonly used measure of performance is the long-run average expected cost

$$\bar{v}(\pi, s) = \liminf_{t \to \infty} E\left[Z_t | \text{initial state is } s\right] / t.$$

Naturally, one seeks the policy that corresponds to the best measure of performance. For example, one might seek the minimum expected discounted cost function:

$$v(s) = \inf_\pi v(\pi, s).$$

Under fairly general circumstances, the minimum expected discounted cost will be the unique solution of the *dynamic programming equation*:

$$v(s) = \inf_{a \in A}\left\{c(s, a) + \int_S dP(s'|s, a) \int_0^\infty e^{-\alpha t} v(s') dF(t|s, s', a)\right\}.$$

Moreover, the value of $a \in A$ as a functionof s that minimizes the right hand side is the optimal action for state s, that is, considering all $s \in S$, this defines the optimal policy or control function. A similar kind of dynamic programming equation is used to compute the optimal policy under the long-run average expected cost criterion.

Besides formulating the semimarkov decision process, perhaps the hardest part of the dynamic programming approach is solving the dynamic programming equation. The theory offers cosiderable help, but these equations are often difficult to solve. Computer methods and approximation techniques are frequently necessary.

Let us now see how to formulate a service process control problem as a semimarkov decision model. Consider an $M/G/1$ queue with arrival rate λ, service distribution G, mean service time $1/\mu$, and $\lambda < \mu$. The server is either present or on vacation. Each time the server goes through a present-vacation-present cycle, the cost K is incurred. The operating cost c per unit time is incurred while the server is present, and the holding cost h is incurred per unit time per customer in the facility.

The problem is to minimize the long-run expected average cost. Intuitively, with few customers in the system the server might be better off being on vacation in order to save on the operating cost. But with a large number of customers, the server should be present in order to reduce the holding cost. And because of K, the server can't be changing status too often. The question is: what do you do in each state?

To formulate this problem as a semimarkov decision model, let the state space S consist of all the pairs $s = (x, n)$, where $x = p$ means the server has been present, $x = v$ means the server has been on vacation, and n is the number of customers in the system. Let $A = \{p, v\}$. We assume action v must be chosen if no customers are present, but otherwise there is a decision to be made. The decision points are the customer arrival epochs when the server is on vacation and the service completion epochs when the server is present.

Suppose the process enters state $s = (v, n)$ with $n \geq 1$. If action $a = v$ is chosen, then the sojourn time will be exponential with parameter λ and the next state will be $s' = (v, n+1)$. Moreover, the cost incurred is nh/λ. On the other hand, if action $a = p$ is chosen, then the sojourn time has distribution G and the next state will be $s' = (p, n-1+N)$, where the random variable N is the number of arrivals during the sojourn time. Also $c(s, a) = nh/\lambda + K + c/\mu + b$, where b is the expected holding cost incurred during the sojourn time by the new arrivals.

For the second main situation, suppose the process enters state $s = (p, n)$ with $n \geq 1$. If action $a = v$ is chosen, then the sojourn time is exponential with parameter λ, the next state is $s' = (v, n+1)$, and the cost incurred is $c(s, a) = nh/\lambda$. On the other hand, if action $a = p$ is chosen, then the sojourn time has distribution G, the next state is $s' = (p, n-1+N)$, and the cost is $c(s, a) = nh/\lambda + c/\mu + b$.

For the third main situation, when the process enters either state $(p, 0)$ or $(v, 0)$ we stipulate that action $a = v$ must be chosen, in which case the sojourn time is exponential with parameter λ, the next state is $(v, 1)$, and the cost is zero.

It is straightforward to compute b and the distribution of N. Then following from the ideas above one can specify the law of motion probability distribution P as well as the sojourn time distribution function F. Next, one writes down and solves the dynamic programming equation for the minimum long-run average expected cost. Finally, one uses this solution to specify the optimal stationary policy. As one might well imagine, this is a long and tedious process, so it will be better to go directly to the final result. It turns out that the optimal policy has a simple form: with Q or more customers, the server should be present, and he should remain present until the number of customers reaches zero; with zero customers, the server should be on vacation, and he should remain on vacation until the number of customers hits Q. Moreover, the integer Q is within one of the quantity

$$\sqrt{\frac{2K\lambda(1 - \lambda/\mu)}{h}}.$$

It is perhaps surprising to discover that the server operating cost c does not appear in this expression. This is because the server will be on the job $1/\mu$ time units per customer, independent of the value of Q. If we allowed the server to be present while the system is empty, then for small enough c it would be optimal for the server to be on job all the time, that is, $Q = 0$.

We conclude this discussion of dynamic optimal control by looking at optimal policies for batch service queueing systems. This is based on the work by Weiss and Pliska (1982). Customers arrive at a batch service system according to a Poisson process. Periodically, all of the customers present are served in one batch, and this service time, which is denoted by S and has a distribution F, does not depend on the size of the batch. Once this service starts, any arriving customers must wait for the next batch. Thus the process $X = \{X_t; \ t \geq 0\}$, where X_t is the number of customers waiting for service at time t, behaves like a Poisson process except that $X_t = 0$ for each time t that the service of batch is initiated.

Two kinds of costs are incurred. Each time a batch is served the cost $K > 0$ is incurred; this cost does not depend on the batch size. Let C_t be the waiting cost rate at time t. If the service of a batch is initiated at time s, then assume C_t is nondecreasing until the next service initiation epoch. In general, the value of C_t may depend on the history of X since the last service initiation epoch. For example, one could have a linear waiting cost $C_t = hX_t$, where h is a positive constant. More generally, one could have $C_t = h(X_t)$ for some nondecreasing function h, so the waiting cost rate per customer varies with the total number waiting. In particular, if h is also convex, then the customers become more irritated as the waiting room becomes more crowded. Finally, for a more complicated example, let f be a nonnegative, nondecreasing function on $(0, \infty)$ and interpret $f(s)$ as the waiting cost rate for a customer who has been waiting s time units. Let $t_i, i = 1, 2, \ldots$, denote the arrival time of the ith customer since the last service initiation epoch, and set

$$C_t = \sum_{i=1}^{X_t} f(t - t_i)$$

(assume this sum is zero if $X_t = 0$). Now C_t is not only a function of X_t, but also of the history of the queue length process since the last service initiation epoch. Since f is nondecreasing, the customers become more irritated the longer they wait.

The problem is to choose a policy for deciding when to serve the batch so as to minimize the average cost per unit time. In particular, one must select what is called a *stopping rule*, which measures the time between successive service initiation epochs. This stopping rule, denoted T, can depend on the history of the queue length process X, and thus on the evolution of the cost rate process C, since the last service initiation epoch. If it is assumed that a service cannot be initiated before the service of the preceeding batch is completed, then it must be that $T \geq S$.

It is important to keep in mind that T is a random variable. For example, suppose

$$T = min\{t \geq S : X_t \geq m\}.$$

This says the service is initiated as soon as the prior service is completed and the number of customers hits the number m, and this is an example of what is called a *control − limit policy*. Or suppose

$$T = min\{t \geq S : C_t \geq k\};$$

this says the service is initiated as soon as the prior service is completed and the waiting cost rate hits the number k.

Once a suitable stopping rule T is adopted, it is used for every service cycle. Hence the service cycles are identical in a probabilistic sense, and the average cost per unit time is simply given by

$$\phi(T) \equiv \left(K + E\left[\int_0^T C_t dt \right] \right) / E[T].$$

In other words, the average cost per unit time is the average cost per cycle divided by the average length of the cycle.

Let ϕ denote the minimum average cost per unit time, that is

$$\phi = \inf_T \phi(T),$$

where the infimum is over all stopping rules T with $T \geq S$. Then, as shown by Weiss and Pliska (1982), the optimal stopping rule T^* is given by

$$T^* = \inf \{ t \geq S : C_t \geq \phi \}.$$

In other words, it is optimal to serve the next batch as soon as the service of the preceeding batch is completed and the waiting cost rate hits ϕ. In particular, if $C_t = h(X_t)$ for some nondecreasing function h, then $T^* = \inf \{ t \geq S : X_t \geq m \}$ for some number m, and in this case a control-limit policy is optimal. On the other hand, if $C_t = \sum f(t - t_i)$ as in the example above, then it may not be possible to express T^* as a control-limit policy.

While this result enables one to characterize the optimal stopping rule, it remains to explain how to compute ϕ and T^*. One can use a simple algorithm if it is assumed that the service time S is instantaneous (this may be reasonable if the service time is relatively short or if there are multiple servers). For each positive number d denote

$$T_d = \inf \{ t \geq 0 : C_t \geq d \}.$$

Define the real-valued function g on $(0, \infty)$ by $g(d) \equiv \phi(T_d)$. Then by Weiss and Pliska (1982),

1. g is minimized by $d = \phi$,
2. ϕ is the unique solution of $g(d) = d$,
3. ϕ is the unique local minimum of g.

Thus ϕ and T^* can be computed by using a simple search technique (for example, Fibonacci search) to find the minimum of g. For example, pick a number d, specify T_d, compute $g(d)$, and check whether $g(d)$ is sufficiently close to d. If not, select another value of d and repeat. During this algorithm, if it is too difficult to explicitly compute

$g(d)$ for particular values of d, then $g(d)$ can nevertheless be estimated with computer simulation, as illustrated by Weiss and Pliska (1982).

5. MANAGEMENT OF QUEUEING NETWORKS

We conclude this study of how to manage queueing systems with a brief look at queueing networks. Prior to this point we have been concerned with systems where the customers wait for a single episode of service and then depart the system as soon as this service is completed. With queueing networks, however, individual customers may proceed through two ore more distinct episodes of service before they depart the system. Items being manufactured by a series of machines, patients in a hospital emergency room, jobs in a computer network, and messages in a telecommunications network are examples of situations that can be modeled as queueing networks.

In order to manage a queueing network, one is faced with the same kind of issues as with ordinary queues. One needs to develop the best design for the system, and then after the system is installed one needs to introduce good dynamic controls. Not surprisingly, one uses the same kind of analytical approaches to analyze these decisions as with ordinary queues: steady-state, equilibrium methods are used for design problems, and dynamic programming techniques are employed for control problems. Only the underlying queueing models are different.

Because of these similarities, and because design issues for queueing networks arise naturally in the study of specific application areas, this section will not attempt to provide a comprehensive survey of how to manage queueing networks. Instead, the focus will be on the dynamic programming methodology for determining optimal controls. Two examples of dynamic, optimal control problems will be presented.

The first example is Weber and Stidham's (1987) study of how to dynamically choose the service rates in certain queueing networks. Although their results can be extended to several kinds of networks, they focus on a cycle of m queues, where a customer finishing queue i proceeds to queue $i+1$, $i = 1, ..., m-1$, and a customer finishing queue m proceeds to the first queue. Customers also arrive at queue i from outside according to a Poisson process at rate λ_i (the special case $\lambda_1 = ... = \lambda_m = 0$ is a closed network). Service at queue i is exponentially distributed at rate μ_i, where μ_i is a decision variable chosen from the interval $[0, \mu]$, where $0 < \mu < \infty$. A cost is charged at the rate $c_i(\mu_i)$ per unit time while the rate μ_i is in effect at queue i, where c is continuous and convex. The number of customers at queue i is denoted by X_i, so $X \equiv (X_1, ..., X_m)$ is the state of the system. The holding cost per unit time is given by $h(X) = h_1(X_1) + ... + h_m(X_m)$, where each h_i is nonnegative and convex. The objective is to choose the m service rates in a dynamic way so as to minimize the expected discounted costs over an infinite planning horizon with discount factor $\alpha > 0$.

To analyze this problem, Weber and Stidham (1987) do a "trick", converting their problem to an equivalent one that is easier to solve. They restrict their decisions to n decision points, namely, n successive epochs that are either customer arrival times or "potential" service completion times. In the transformed model the server at queue i is always running at the maximum rate μ, but when a potential service completion occurs, it is "accepted" (i.e., the customer actually departs) with probability μ_i/μ, whereas it is "rejected" (i.e., recycled at the same queue) with probability $1 - \mu_i/\mu$. As a result, the time between successive decision points will be exponentially distributed with parameter $\Lambda \equiv m\mu + \lambda_1 + ... + \lambda_m$.

Since service rates need only be changed at decision points, the dynamic programming equation takes a more simple form. Let e_i denote the m-vector which has 1 in component i and 0 elsewhere; let $d_i \equiv e_{i+1} - e_i$, $i = 1$, ..., $m - 1$; and let $d_m = e_1 - e_m$. Suppose X is the current state and μ_i is the chosen service rate at node i, $i = 1$, ..., m. Consider the state at the next decision point. With probability λ_1/Λ the next decision point is an arrival to queue i, in which case the system moves to the state $X + e_i$. With probability μ_i/Λ the next decision point is an actual service completion at queue i, in which case the system moves to state $X + d_i$. Finally, with probability $(m\mu - \sum \mu_i)/\Lambda$ the next decision point is a null event, in which case the system remains in state X. Let $V_n(X)$ denote the minimum expected discounted value of the costs starting from state X and running for n decision points. Then by standard dynamic programming theory (see Heyman and Sobel (1982, vol. 2) or Whittle (1982)), V_n satisfies the functional equation

$$V_{n+1}(X) = \sum_{i=1}^{m} \min_{\mu_i}[h_i(X_i) + c_i(\mu_i) + \lambda_i V_n(X + e_i) + \mu_i V_n(X + d_i)$$

$$+ (\mu - \mu_i)V_n(X)]/(\alpha - \Lambda).$$

With $V_0(X) = 0$, this equation can be solved in a recursive manner.

Letting $n \to \infty$, the sequence V_n converges to V, the minimum expected discounted cost over an infinite planning horizon. This V satisfies the same equation, only with V substituted for both V_n and V_{n+1}. Let $\mu_i(X)$ be the value of μ_i that minimizes the right hand side with V in place of V_{n+1}. Then $\mu(X) \equiv (\mu_1(X), ..., \mu_m(X))$ is the vector of optimal service rates as a function of the state.

Weber and Stidham did not make any general conclusions concerning the difference $\mu(X + e_i) - \mu(X)$, that is, how the optimal service rate vector changes as a single customer is added to queue i. However, by analyzing the dynamic programming equation they were able to prove that

$$\mu_j(X) \leq \mu_j(X + d_i)$$

for all $j \neq i$, and

$$\mu_i(X) \geq \mu_i(X + d_i) .$$

In other words, as a customer moves from queue i to $i + 1$, the optimal service rate at queue i does not go up and the other $m - 1$ optimal service rates do not go down.

For our final example of a dynamic control problem for a queueing network, consider the case where there is a single server or operator for the whole network and the problem is to decide at which node the operator should be working at each moment in time. This is the problem analyzed by Tcha and Pliska (1977).

Customers arrive at each of m nodes from outside the system according to a Poisson process with rate λ_i, $i = 1, ..., m$. Each time a customer's service at station i is completed, with probability p_{ij} he proceeds to station j or with probability $p_{i0} \equiv 1 - p_{i1} - ... - p_{im}$ he leaves the system. The service time at node i has distribution function F_i. The system incurs the cost h_i per unit time for each customer at node i, but receives the reward r_{ij} when a customer departs i and proceeds to node j, $j = 0, 1, ..., m$ ($j = 0$ means the customer departs the system).

While the single operator is working at one node, all the customers at other nodes are waiting. The problem is to schedule this operator so as to maximize the expected discounted reward over an infinite planning horizon. This schedule will take the form of a rule that specifies at any decision point where the operator should be as a function of the state of the system, that is, the distribution of the customers among the m nodes. The *decision points* are certain epochs between which the operator must remain at the same node. Two cases are considered. First is the *non−preemptive case*, where the decision points are the service completion times plus those epochs when a customer arrives to find the operator idle. In this case the service times are general, nonnegative random variables satisfying $F_i(0) < 1$. In the other case, called the *preemptive case*, the decision points are all the service completion times plus all the customer arrival times, but now the service times are exponential random variables.

This queueing network control model is useful for several applications where the nodes are not different locations but rather are different kinds or categories of customers. For example, consider a single-operator repair facility where jobs arrive and proceed from one category to another according to what inspections, tests, and repairs are performed. Or consider a computer system where jobs change categories as executing proceeds. Finally, consider a single-technician medical laboratory in which samples arrive to be processed through different sequences of tests.

It turns out that for both cases a *modified static policy* is optimal. A modified static policy has a very simple form: for some integer K (with $0 \le K \le m$) and some ordering of the nodes, customers at nodes $K + 1$ through m are never served. At any decision point, customers at node 1 are served first; if none is present there, then customers at node 2 are served first; and so forth through node K.

The analysis of this model is carried out with the methods of dynamic programming. First, for an arbitrary modified static policy, a formula is derived for the expected discounted return as a function of the number of customers at each node. Second, this formula and dynamic programming theory are used to establish a condition sufficient to imply that a particular modified static policy is optimal among all possible policies. Finally, an algorithm is developed for computing the modified static policy that satisfies this sufficient condition and thus is optimal. The details of this algorithm, which differs slightly for the two cases, are too long to report here; the interested reader should consult Tcha and Pliska (1977).

6. REFERENCES

1. Bell, C.E. 1980. Optimal Operation of an $M/M/2$ Queue with Removable Servers. *Opns. Res.* **28**, 1189-1204.

2. Chen, H., et al. 1988. Empirical Evaluation of a Queueing Network Model for Semiconductor Wafer Fabrication. *Opns. Res.* **36**, 205-215.

3. Conway, R.W., Maxwell, W.L., and Miller, L.W. 1967. *Theory of Scheduling*. Addison-Wesley, Reading, Massachusettes.

4. Crabill, T.B., Gross, D., and Magazine, M.J. 1977. A Classified Bibliography of Research on Optimal Design and Control of Queues. *Opns. Res.* **25**, 219-232.

5. Heyman, D.P., and Sobel,M.J. 1982. *Stochastic Models in Operations Research*. Vol.1 and 2. McGraw-Hill, New York.

6. Jaiswal, N.K. 1968. *Priority Queues*. Academic Press, New York.

7. Lu, F.V., and Serfozo, R.F. 1984. $M/M/1$ Queueing Decision Processes with Monotone Hysteretic Optimal Policies. *Opns. Res.* **32**, 1116-1132.

8. Ohno, K., and Ichiki, K. 1987. Computing Optimal Policies for Controlled Tandem Queueing Systems. *Opns. Res.* **35**, 121-126.

9. Powell, W.B., and Humblet,P. 1986. The Bulk Service Queue with a General Control Strategy: Theoretical Analysis and a New Computational Procedure. *Opns. Res.* **34**, 267-275.

10. Rothkopf, M.H., and Rech, P. 1987. Perspectives on Queues: Combining Queues is not Always Beneficial. *Opns. Res.* **35**, 906-909.

11. Sobel, M.J. 1974. Optimal Operation of Queues. *Lecture Notes in Economics and Mathematical Systems*. No.98, Springer-Verlag, Berlin, 231-261.

12. Sobel, M.J. 1982. The Optimality of Full Service Policies. *Opns. Res.* **30**, 636-649.

13. Stidham, S. 1970. On the Optimality of Single-Server Queueing Systems. *Opns. Res.* **18**, 708-732.

14. Stidham, S., and Prabhu, N.U. 1974. Optimal Control of Queueing Systems. *Lecture Notes in Economics and Mathematical Sciences*. No.98, Springer-Verlag, Berlin, 263-294.

15. Szarkowicz, D.S., and Knowles, T.W. 1985. Optimal Control of an $M/M/s$ Queueing System. *Opns. Res.* **33**, 644-660.

16. Sze, D.Y. 1984. A Queueing Model for Telephone Operator Staffing. *Opns. Res.* **32**, 229-249.

17. Tcha, D.W., and Pliska, S.R. 1977. Optimal Control of Single-Server Queueing Networks and Multi-class $M/G/1$ Queues with Feedback. *Opns. Res.* **25**, 248-258.

18. Weber, R.R., and Stidham, S. 1987. Optimal Control of Service Rates in Networks of Queues. *Adv. Appl. Prob* **19**, 202-218.

19. Weiss, H.J., and Pliska, S.R. 1982. Optimal Policies for Batch Service Queueing Systems. *OPSEARCH* **19**, 12-22.

20. Whittle, P. 1982. *Optimization Over Time.* Wiley, New York.

21. Yao, D.D. 1987. The Arrangement of Servers in an Ordered-Entry System. *Opns. Res.* **35**, 759-763.

Chapter 10
QUEUEING NETWORK MODELS
FOR
FLEXIBLE MANUFACTURING SYSTEMS

N. Erkip & S. Meral
Department of Industrial Engineering
Middle East Technical University
Ankara, Turkey

1. INTRODUCTION

1.1. Definition of Flexible Manufacturing System

A flexible manufacturing system (FMS) consists of a group of processing stations, usually numerical control (NC) machines, and associated storage elements, connected together by an automated workpart handling system. It is controlled by a computer or a network of computers. The purpose of the flexibility and versatility of the configuration is to meet production targets for a variety of part types in the face of disruptions such as demand variations and machine failures. Human labor is also used to carry out some functions to support the operation of the FMS, as loading raw workparts onto the system, unloading finished parts from the system, changing tools, etc.; in addition to these, the operators may be required to input data, change part programs, and perform other tasks related to the computer control system.

1.2. Problems Solved by FMS

FMSs are designed to fill the gap between high-production transfer lines and low-production NC machines. The relative position of the FMS concept is illustrated in Figure 1. Low-production volumes are encountered in job-shops whereas high-production volumes are in flow-shops. An FMS is designed to have "medium" production volumes and is comparable to Group Technology's (GT) production amounts. FMS is not a new technology but intends to utilize the already existing technology more efficiently. It can be applied to the production of medium-volume batches of various sizes in a number of industrial sectors, such as metal working industries.

The FMS concept is intended to solve the following application problems which are inherent in the traditional manufacturing systems (Groover (1980)) :

188

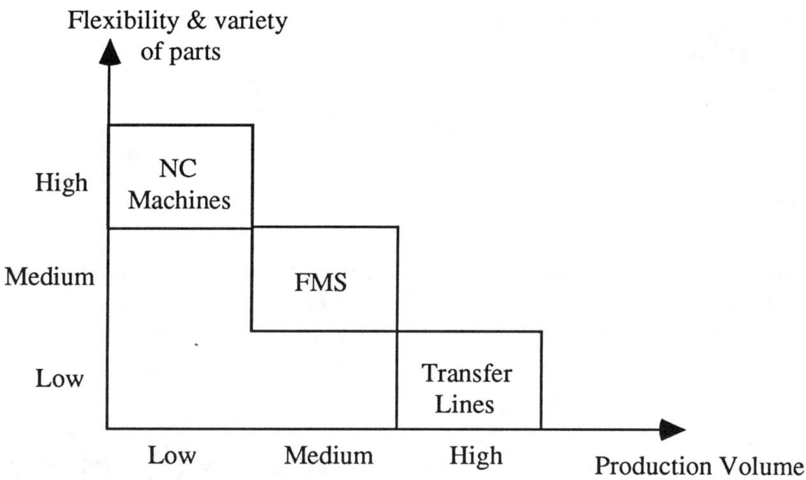

Figure 1. Relative Position of the FMS Concept

1. Production of families of workparts,
2. Random launching of workparts onto the system,
3. Reduced manufacturing lead time,
4. Reduced in-process inventory,
5. Increased machine utilization,
6. Reduced direct and indirect labor,
7. Better management control.

Most of the outlined problems are not new, but high FMS investment costs and automation require a new approach to the solution of these problems.

Machine tools, material handling system, storage areas for in-process inventory, and computer control are the FMS components usually used to describe and classify different types of FMSs.

An overview of the FMS concept and description of FMS can be found in Bessant (1984), Black (1983), Browne et al. (1984), Cook (1975), Curtin (1983), Groover (1980), Keywood (1983), Jablonowski (1985), Suri and Whitney (1984).

1.3. Outline of the Chapter

Before discussing the queueing network models for FMSs, transfer lines with a relatively higher production rate but a lower flexibility with respect to FMSs are discussed in Section 2, together with the available techniques to find the steady state probabilities and the optimal configuration of buffers.

In Section 3, it is shown that FMSs with more complex structures than transfer lines can be represented as networks of queues for the purpose of performance evaluation and design issues.

In Section 4, the theory of queueing network models in relation to FMS is discussed in four categories with their current solvabilities.

Section 5 surveys an overview of the models, both evaluative and generative, specially designed for FMSs. Section 6 is the conclusion.

2. TRANSFER LINES

Transfer line is a production system that has a number of stations connected in series where all workparts move from one station to the next and rest between stations in buffers. A schematic diagram of such a system appears in Figure 2 below. A transfer line usually has a higher production rate and a lower flexibility than FMS.

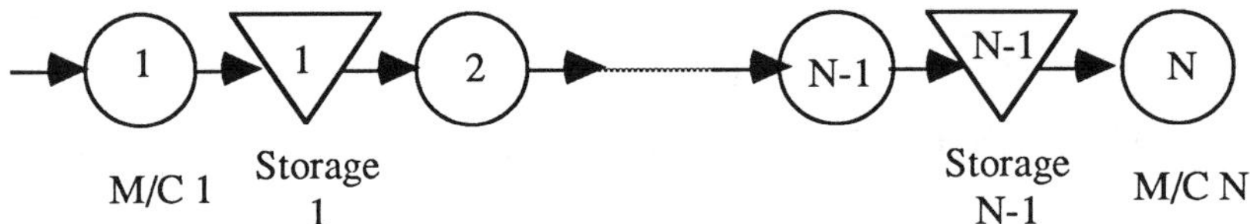

Figure 2. An N-Stage Transfer Line

A problem often encountered in justifying the installation of an automatic transfer line is the low efficiency of the line compared to the use of individual machines (Buzacott (1967)).

2.1. Buffer Storage Design Problem

The machines in a transfer line may fail at random times and remain inoperable for random periods while they are under repair. So, each machine is in one of the following two states: operational or under repair (or failed). In addition, a machine can be blocked or starved. A machine is blocked if the service of a part on this machine is completed and the next buffer is full, therefore it is not possible for the part to enter the buffer. In this case, the machine remains idle until service in the next machine is completed. A machine is starved if there is no part in service or the buffer feeding the machine is empty. When a machine is blocked or starved, even if it is operational, it cannot operate; therefore it cannot fail.

It is possible to compensate for machine failures by providing redundancy, i.e., parallel machines are brought into use in case of failures of primary machines. This process, however, can be prohibitively expensive, if system components are costly. An alternative is to place buffer storages between unreliable stages. These provide temporary storage space for the products of stations upstream of a failed station. Thus, they decrease the effects of work station failures on the rest of the line. However, cost of keeping buffer storages, i.e., sum of costs of floor space, material handling equipment, and in-process inventory are also important and should be considered. It is thus necessary to find the "best" storage configuration; this leads to an optimization problem.

2.2. Analysis of Unreliable Transfer Lines Using Markov Processes

To represent the "best" storage configuration, the design parameters and the performance measures should be represented in terms of the steady state probabilities. The design parameters are the average up and down times of the workstations and sizes of the buffers between stages.

The states of an isolated machine, namely operational and under repair, constitute a Markov chain with given failure and repair transition probabilities:

q = Probability {Machine down next cycle | up this cycle },
a = Probability { Machine up (repaired) next cycle | down this cycle }.

Buzacott (1967) shows via Markov analysis that

Mean Time Between Failures = $1/q$,

Mean Time to Repair = $1/a$,

and the stand-alone efficiency of the machine is

$$\tau = \frac{1}{1 + q/a}$$

In the extreme case of N connected machines without a buffer, failure of any one machine will result in the "force down" of the entire production line (Buzacott (1967), Ho et al. (1979)).

A similar Markov analysis quickly shows that the efficiency of the zero-buffer line is:

$$\tau_0 = \frac{1}{1 + \sum_{i=1}^{N} q_i/a_i} < \frac{1}{1 + q_i/a_i} \equiv \tau_i$$

for all i where q_i and a_i are the parameters of machine i. At the other extreme, if an infinite amount of buffer is provided between all two consecutive machines for intermediate storage of semi-finished pieces, then the production efficiency of such a line is bounded from above by $\tau_\infty \leq \min_i[\tau_i]$ as the slowest stage will determine the efficiency. For finite amount of buffer storages, the production line efficiency,τ, is bounded from above and below by τ_∞ and τ_0, respectively.

The buffer storage design problem can then be stated as one of the types below:

1. Choose the size of buffer storages $b_1, b_2, \ldots, b_{N-1}$ to maximize τ for given $\{a_i, q_i\}$; $\{i = 1, ..., N\}$ subject to cost constraints (Ho et al. (1979)).
2. Minimize total design cost (or maximize profit) subject to a minimum acceptable level of τ.

In the general case where cycles can not be specified, we need to use continuous-time Markov chains. It is clear that for finite buffer storages $b_1, b_2, ..., b_{N-1}$, the states of the production line constitute a Markov process. Each machine can be up or down; each buffer can have up to b_i parts, including zero. The total number of states is

$$N = 2^N \prod_{i=1}^{N-1} (b_i + 1).$$

The corresponding transition rate matrix A can be specified accordingly in terms of the various a_i's and q_i's. Note that a_i's and q_i's are rates in this case. Once this is done, steady state probability of state i, π_i, $i = 1, ..., N$ in the Markov process can, in principle, be determined via the solution of the equation $\pi A = 0$. Production efficiency of the line is then simply the sum of the π_i's, $i \in A_0$, where A_0 is the set of output states, i.e., states that result in a part being produced by the line. Thus, conceptually line efficiency or throughput τ can be expressed as a function of all b_i's parameterized by the given a_i's and q_i's.

Unreliable two-stage systems with finite buffers have been studied by many researchers. Longer systems are more difficult to analyze because of the complexity of machine interference when buffers are full or empty. When both breakdowns and blocking are included, complexity of the model increase substantially. Thus, efficient approaches are required to compute the steady state probabilities.

2.3. Approaches for Longer Transfer Lines

The method of Gershwin and Schick (1983) which exploits the special structure of the related Markov processes, can be extended to longer lines, more complex networks, and other models of machines and buffers (although it has only been applied to two and three-stage systems). The main assumption is that processing times at different stages are equal.

Their approach can be seen as a general technique for computing the steady state probabilities for large scale, structured Markov process problems. Unfortunately, Gershwin (1987b) reports unsuccessful computational experiences.

Gradient technique proposed by Ho et al. (1979) seems to be a more systematic approach for buffer sizing. This technique computes the sensitivity (gradient) of the increase of the transfer line production per unit buffer size increase at each location. The method of computing is coupled with a simulation model resulting in an efficient estimation of the transfer line efficiency for various buffer configurations. Using the gradient technique, solution to large design problems of type (1) in Section 2.2 can be found for problems that accommodate arbitrary machine failure and repair time distributions.

An efficient decomposition method, presented by Gershwin (1987a), for the approximate evaluation of tandem queues with finite storage space and blocking is applicable for both reliable and unreliable tandem systems where processing times at all stages are equal. For the three-machine case, the approximation method produces results that are extremely close to the exact values obtained by solving the Markov chain. Exact methods are not available for systems of more than three machines and two buffers or for three-machine cases with very large buffers. Consequently, other techniques are required to assess the accuracy of the approximation method. These include simulation and qualitative observations. The most important conclusion related to this approach is that there is a close agreement between the approximation and the simulation results. This remains true for large buffer capacities (over 100) and long lines (20 machines). There is no obvious trend indicating that the accuracy of the approximation decrases as the number of machines increases.

Another approximation method to analyze tandem queues with blocking is proposed by Brandwajn and Jow (1988). The definition of blocking is a bit different than the one discussed previously in this paper: a server is not allowed to start service until space is available in the downstream buffer. The method proposed by Brandwajn and Jow (1988) can handle both types of blocking, but they carried out their work on the one defined above. The approach is based on the equivalence of the tandem system to a specified two-node system. It uses marginal probability distributions together with an approximate evaluation of the conditional probabilities introduced through such an equivalence. In the equivalent two-node system, the two nodes have the same limited capacity as the original system. The equivalence results in an approximate representation of interarrival and service times at the second node as exponentially distibuted random variables, and the service at the first node is executable only when the second queue is full. Once we know how to solve a two-node system, the "equivalent" rates of arrival and service become, approximately, the service rate and the arrival rate for the preceding and succeeding (with respect to the two-node system) nodes, respectively. As a result of this approximation, an iterative approach that considers the nodes in the original system is developed. The method can not be proved to converge, but limited experiments indicate that it fastly converges. The numerical results given seem to be very promising. The method can also be extended to tandem queues with state dependent service rates which is an issue not covered by existing approaches.

Another study for long transfer lines has been carried out by Altıok and Stidham (1983) to find the optimal allocation of buffer capacities so that the average profit is maximized. The queueing system is transformed to an equivalent Markovian system to find the steady state probability distributions. The general network is represented as a network in which each station is of Coxian type with a limited queueing capacity. This network with a phase-type state definition forms a Markov process. The advantage of Coxian formulation is that it can also be used as an exact or approximate representation for systems with nonexponential service times, with or without breakdowns. Coxian service time distributions and finite buffer capacities can be studied through decomposition approximations. An approximate decomposition procedure is presented by Altıok

(1985). This procedure has two levels of approximations. The first one is the moment approximation of the service completion times by phase-type random variables and the second is the decomposition of the resultant system into individual stations with revised arrival and service processes. The approach performs better in balanced production lines, i.e., when the utilizations of stages are nearly equal.

In most of the models discussed above, the processing times of parts at each stage are equal at all machines. Gershwin (1987b) extends their previous technique of decomposition to systems in which processing times at different machines are allowed to take different values. It is not a new algorithm; rather it is a way of representing machines of different speeds. Numerical and simulation experiments suggest that the method is effective; although analytic results on bounds of the approximation are not available yet.

2.4. Some Remarks on Transfer Lines

Research on transfer lines mainly concentrates on two aspects:

1. To find efficient approaches to compute the steady state probabilities of the line which allows for unreliable machines, finite buffers, starvation and blocking,
2. To find the optimal configuration of buffers.

Especially the latter is not a straightforward task and should take into account the difficulty of computing the probabilities.

3. REPRESENTATION OF FMS AS A NETWORK OF QUEUES

3.1. Definitions

Unlike a transfer line where all workparts follow a sequential route through the system, an FMS permits workparts to visit work stations in any arbitrary sequence.

A work station or a machining center that processes workparts is called a **resource**. The resource may have one or more identical machines, called **servers**. The items or workparts are called **jobs**. A **queueing network** (open or closed) consists of individual queues corresponding to resources in the manufacturing system that are interconnected by the flow of jobs through various resources. An FMS can be modeled as a queueing network because such systems are primarily affected by contention for resources, resulting in congestion of flow through the system. A queueing network model of an FMS consists of a finite number of distinct resources, each with an associated queue and a queue discipline, usually FCFS, where jobs may wait prior to being processed. Figure 3 illustrates a model of an FMS as a closed network of queues. Each resource may have one or more identical servers, and jobs circulate in the network. A job, after being processed at resource i, proceeds to some resource j.

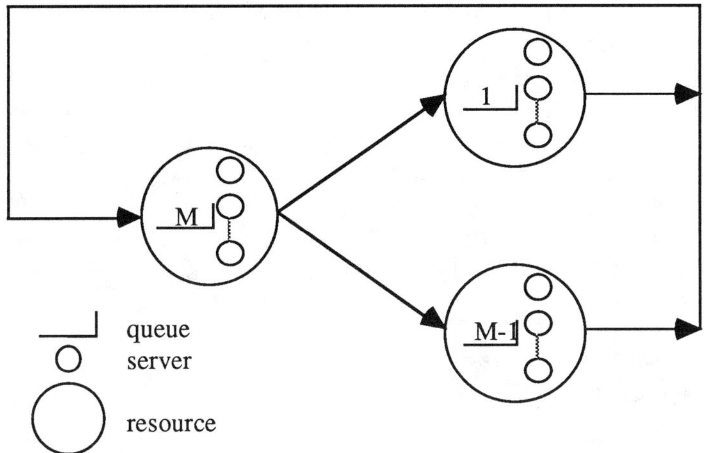

Figure 3. A Closed Queueing Network of a FMS with M Resources

3.2. Queueing Network Formulation as a Decision Aid for FMS

Queueing network models can be effectively used as performance models of FMSs and also as an aid in design issues. A performance model is used to evaluate and predict the stochastic behavior of work flow in the system to aid decision making. By capturing the stochastic behavior of work flow through the system, a number of useful performance measures can be determined as throughputs, queue length distributions, utilizations, and mean waiting times.

Recently, a set of arguments presented by Suri (1983) explain the robustness of analytical queueing network models for the representation of practical systems. Some performance measures such as system throughput and machine utilization have been shown to be extremely robust to violations in the assumption of homogeneous service times. This assumption can also be thought as the only restrictive assumption in modeling many systems. So, these results justify the use of queueing network formulas for the modeling and optimization of several practical systems, including FMSs , which are likely to violate the assumptions of classical queueing network theory. However, one should be extremely careful in utilizing queueing network models as decision aids in FMS's. The assumptions behind queueing network models and their implications should be well comprehended. Co and Wysk (1986) consider the well known CAN-Q model by Solberg (1977) and examine the effects of assumptions. They come up with a set of conclusions related to the use of such a model. A preliminary analysis of this sort will be very helpful before applying a queueing network model of FMS.

In the following sections, networks of queues and the representation of FMSs as networks of queues will be discussed.

4. OVERVIEW OF NETWORKS OF QUEUES AS RELEVANT TO FMS

In a closed network, the number of customers is constant since customers can neither enter nor leave the system, but pass repeatedly through various stages. On the other hand, in an open system, at any instant, the number of customers is a random variable since customers that enter the system are serviced at various stages and exit (Gordon and Newell (1967)). In a queueing network, the queue lengths may be either restricted or unrestricted. So, depending on whether the total number of customers in a network is fixed and on whether the maximum queue length at each node is restricted, queueing networks can be works can be categorized as:

1. Open unrestricted,
2. Closed unrestricted,
3. Closed restricted,
4. Open restricted.

4.1. Unrestricted Networks of Queues

4.1.1. Open Unrestricted Networks of Queues

In his pioneering work, Jackson (1957) derives analytical formulas for the performance parameters of an open network in terms of the parameters of servers (machines). The characteristics of the machine shop studied are:

1. Machine shop has several departments (M), each containing a fixed number of identical machines,
2. The waiting jobs are pooled in a single line in front of each machine,
3. Service times are exponentially distributed with parameter μ_m for department m,
4. Arrivals at a given department come both from other departments and from outside the shop. The number of customers in the system is countably infinite,
5. Those coming from outside to any department m arrive in a Poisson-type time series at a mean rate λ_m,
6. The flow pattern of jobs inside the shop can be described of follows:

When a department finishes a job, the job either goes to some specified department or leaves the system. Its particular course is governed by a fixed probability distribution (after being served in department m, the job goes to department k with probability θ_{mk}, the proportion of jobs that go from m to k).

If mean arrival rates at various departments (say Γ_m) are properly defined, then the result is a steady state distribution in which the queue lengths of the departments are independent.

If Γ_m is the average arrival rate of customers at department m from any source, inside or outside the system, then in the steady state

$$\Gamma_m = \lambda_m + \sum_k \theta_{km}\Gamma_k \; , \qquad m = 1, ..., M$$

If n_m represents the number of customers waiting and in service at department m and r_m represents the number of machines (servers) at department m, then the probability that there are n_m customers at department m is

$$P^m_{n_m} = \begin{cases} \left[(\Gamma_m/\mu_m)^{n_m} / n_m! \right] P^m_0 & , \; 0 \leq n_m \leq r_m \\ \left[(\Gamma_m/\mu_m)^{r_m} / r_m! r_m^{n_m - r_m} \right] P^m_0 & , \; n_m \geq r_m \end{cases}$$

where P^m_0 is obtained after normalizing all probabilities.

The steady state distribution of the system is given by the product

$$P(n_1, n_2, ..., n_M) = P^1_{n_1} \cdot P^2_{n_2} \cdots P^M_{n_M}$$

provided that $\Gamma_m < \mu_m r_m$ for $m = 1, ..., M$. This system behaves as if its departments were independent elementary systems.

4.1.2. Closed Unrestricted Networks of Queues

Jackson (1963), which is an extension of his earlier work discussed above, Jackson (1957), provides the equilibrium joint probability distribution of the queue lengths for a broader class of queueing-theoretical models. In the model studied, arrival rates and service rates are state dependent, that is, the probability that a customer will actually enter the system for service depends on the present number of customers, and the mean service rate at each service center depends on the queue length at each service center.

Gordon and Newell (1967) obtained the equilibrium equations for the joint probability distribution of customers as done in Jackson (1963), but solved them by a "separation of variables" technique, obtaining a simplified notational structure.

The unique equilibrium distribution for a closed system with N customers is given by

$$P(n_1, n_2, ..., n_M) = \{\prod_{i=1}^{M} \left[\frac{x_i^{n_i}}{\beta_i(n_i)} \right] \} G^{-1}(N)$$

where n_i is the number of customers at stage i, r_i is the number of parallel servers at station i, $N = \sum_{i=1}^{M} n_i$ is the allowable number of customers in the closed network,

$$\beta_i(n_i) = \begin{cases} n_i! & , \; n_i \leq r_i \\ r_i! r_i^{n_i - r_i} & , \; n_i > r_i \end{cases}$$

and $G(N)$ is the function determined from the normalization

$$G(N) = \sum_{\sum_{i=1}^{M} n_i = N} \left\{ \prod_{i=1}^{M} \left[\frac{x_i^{n_i}}{\beta_i(n_i)} \right] \right\} .$$

Here x_i's are parameters that are obtained by solving a system of linear equations.

They also showed that closed networks are stochastically equivalent to open systems in which the number of customers cannot exceed N. They examined that as $N \to \infty$, the system is regulated by the stages with the slowest service rate, i.e., bottleneck stages.

Buzen (1973) examines computational aspects of the basic equilibrium distributions in closed queueing networks as well as certain marginal distributions which can be derived from them. The computational algorithms are based on two-dimensional iterative techniques which are highly efficient and quite simple to implement.

In Towsley (1980), a model of closed queueing network is presented within which customer routing between queues may depend on the state of the network. The probability of a transition from queue i to queue j would be a function of the number of customers in several queues. If a network with no state-dependent routing has a product-form joint equilibrium distribution of the queue lengths, then the introduction of the routing functions preserve the product-form of the equilibrium distribution. Also, computational procedures to facilitate the analysis of such complex networks are provided.

According to Koenigsberg (1982), Posner and Bernholtz studied the problem of cyclic queues with time lags. In cyclic queue models, the machine groups are considered fixed and the workparts or jobs move through the production line in a cyclic order. A time lag is defined as the time for a customer to move from one stage to another. Their results are of major importance in understanding the behavior of queueing networks. They proved that the state probabilities are independent of the time lag distribution. Further, they showed that all time lag stages can be lumped into a single "lag" stage, with average time equal to the sum of the individual average time lags, thus effectively adding a single stage to account for all time lags. Except this case, it is usually difficult to deal with transportation times between stations, so in most of the FMS applications the transportation times are neglected.

All of these formulas for both open and closed networks are generalized in a comprehensive paper by Baskett et al. (1975). They extend the results to include different queueing disciplines, multiple classes of jobs and nonexponential service time distributions. Each customer belongs to a single class of customers while awaiting or receiving service at a service center, but may change classes and service centers according to fixed probabilities at the completion of a service request. For open networks, state dependent arrival processes are considered. A network may be closed with respect to some classes of customers, and open with respect to other classes of customers.

4.2. Restricted Networks of Queues

In this case, blocking may occur due to restrictions on each queue length, thereby decreasing the utilization. Blocking occurs when a customer, having completed service at station i, cannot proceed to the succeeding station j because facilities at station j are completely filled. In this case, we say that station i is blocked by station j. Except the stations where the arcs are directed from the system, each station has a positive probability of blocking jobs in the preceding stations and/or having jobs that are blocked by the succeeding stations.

4.2.1. Open Restricted Networks of Queues

The existing methods are quite difficult to apply for solving the steady state probabilities of an open restricted network of queues. The exact solution of the open restricted queueing network is difficult as the number of possible states that should be considered is very large. The number of unknown variables is approximately $2^N \prod_{i=1}^{K} S_i$ in the worst case where S_i, K and N are the size of holding facilities at node i, the total number of nodes in the network, and the total number of arcs, respectively (Takahashi et al. (1980)).

In the study of Takahashi et al. (1980), an approximation method for analyzing open restricted queueing networks with exponential interarrival and service times is presented. The analysis is done with a node-by-node decomposition by introducing pseudo-arrival rates and effective service rates. It is assumed that the service rates can be modified to take into account the possibility of blocking and the effective service rate can be found for each node. Assuming that the effective service time of node i is still exponentially distributed, the approximation proceeds by relating the effective service time of node j to node i where node i succeeds node j. A node-by-node decomposition is used to approximate the open restricted queueing network. This method of analysis makes it possible to obtain easily computed approximations of various characteristic quantities such as blocking probabilities, output rate, etc., for the network.

Other approximation techniques exist for open restricted queueing networks which can be applied in various different circumstances. Yao (1983) considers an FMS as an open restricted queueing network and an iterative scheme is developed for the analysis. The detailed description of the method is given in the next Section.

Schweitzer (1977) studied on open restricted queueing network where blocking is not considered. The study seeks to find the maximum throughput under the assumption that the system is under independent balance. Independent balance equations are sufficient conditions for global balance but they are not necessary. It assumes that rate of flow into a state due to a customer entering a stage of service is equal to the flow out of that state due to a customer leaving that stage of service. The importance of using local balance equations are:

1. It has product-form solutions for equilibrium probabilities,

2. It works for a large number of cases, except FCFS scheduling at a service center with different classes of customers having different service time distributions .

Schweitzer (1977) proves that if service rates are constant for each customer class, and excess arrivals are discarded, then the following two results hold: (1) The maximum possible output rate for the finite-capacity case equals the value of the customer arrival rate which just saturates the slowest server in the infinite-capacity case; (2) An infinite-capacity network can have a strictly greater output rate than the corresponding finite-capacity network, no matter how larger the capacity is.

4.2.2. Closed Restricted Networks of Queues

Most of the studies about closed restricted queueing networks are related to the design issues of FMSs. Methods of Suri and Diehl (1986), Yao and Buzacott (1986a), Yao and Buzacott (1985), and others will be discussed in the next section. Perros et al. (1985) consider a closed restricted queueing network where blocking and deadlock may occur. It is assumed that deadlocks are detected and solved instantaneously. Each queue is served by a single exponential server. The method has three steps:

1. Generation of the state space,
2. Generation of the rate matrix (Perros et al. (1985) point that this is the most crucial step where the effects of dimensionality are seen),
3. Numerical solution technique.

The method is exact and used to approximate the solution for the case where some of the queues are restricted (i.e., the others are unrestricted). The whole network can be divided into two subnetworks: the blocking subnetwork, i.e., group of all finite and infinite queues that are liable to being blocked, and the non-blocking subnetwork, the remaining queues. The exact procedure is used to approximate the systems by forming composite queues, i.e., a subnetwork represented by a queue. Numerical results presented seem to be promising.

Examining the queueing network models in the four categories discussed above, with their current solvabilities, it seems quite reasonable to represent FMSs as closed or open, but restricted networks of queues. The queue lengths or buffer storage capacities are limited above by the capacity of materials handling system or available space. On the other hand, FMS systems, in terms of the number of jobs circulating in the system, may be open or closed. Some of the researchers model FMSs as closed networks of queues because the number of pallets available to transport parts within the system is finite. During the process, all pallets can be occupied and result in a closed system in terms of the number of parts in the system.

4.3. Mean-Value Analysis

Mean-value analysis (MVA) is a rather recent development in networks of queues theory that was first presented by Reiser and Lavenberg (1980) (see also Zahorjan et al. (1988)). It is a technique for computing performance measures as throughputs, utilizations, and mean queue lengths in unrestricted closed networks of queues. It takes its name from the fact that it deals mainly with the first moment (mean value) of statistical distributions associated with the problem.

The traditional approach to the solution of Markovian queueing networks is to formulate a system of balance equations for the joint probability distribution of the vector-valued system state; and the solution of the balance equations is in the form of a product of simple terms; the product terms are to be normalized to form a proper probability distribution. For practical purposes, however, the joint distribution contains far too much detail. Much simpler quantities such as mean queue sizes, mean waiting times, utilizations, and throughputs are needed. Within the framework of the convolution algorithm of Buzen (1973), it has been shown that such quantities can be derived from the normalization constants in an efficient manner. The mean-value algorithm works directly with the desired statistics. Its operations count is said to be asymptotically equivalent to earlier algorithms. The mean-value analysis is based on a relation between the mean waiting time and the mean queue size of a system with one customer less. This mean-value equation, augmented by Little's equation applied to each routing chain and separately to each service center, will furnish a set of equations which are easily solved numerically. The normalization constants are never computed. The new algorithms are simple and avoid overflow/underflow problems which may arise with the convolution algorithm. All mean values for the queue sizes are calculated in parallel.

5. DESIGN AND PERFORMANCE EVALUATION OF FMSs

Since an FMS is a complex system, decision making in the context of FMS can be difficult. Thus we have, on one hand, the complexity of decision making, yet on the other hand, the requirement to design and operate these systems efficiently in order to maximize the utilization of the expensive equipment. A number of recent reviews are available for analytical models of FMSs (see Buzacott and Yao (1986), Kusiak (1986), Von Looveren et al. (1986), Stecke (1986), Stecke (1988), and Erke and Erkip (1986)).

Typical decisions to be made during the planning and initial design phases include:

1. Buffer size,
2. Number of jobs allowed in the system,
3. Resource allocation (allocation of fixed number of buffer space, pallets available, spare tools, etc.),
4. Size of machine groups,
5. Load balance of the machines (machine groups),
6. Part-mix ratio decisions.

Typical performance measures used to evaluate these decisions include :

1. Throughput time,

2. Production capacity (Throughput time asymptotically approaches to production capacity),
3. Mean number of jobs in the system,
4. Mean waiting time in the system.

In considering some decision aids for the FMS life cycle, it is important to differentiate between **generative** models which find good candidate decisions, and **evaluative** models which evaluate a given set of decisions, as seen in Figure 4 below.

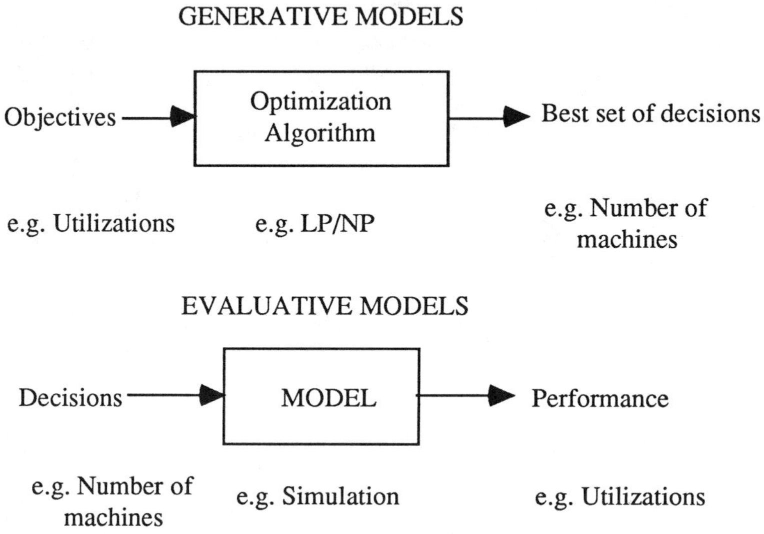

GENERATIVE MODELS

Objectives ⟶ [Optimization Algorithm] ⟶ Best set of decisions

e.g. Utilizations   e.g. LP/NP   e.g. Number of machines

EVALUATIVE MODELS

Decisions ⟶ [MODEL] ⟶ Performance

e.g. Number of machines   e.g. Simulation   e.g. Utilizations

Figure 4. Classification of Decision Aids by Suri (1984)

5.1. Evaluative Models for FMSs

Main models to evaluate the performance of FMSs are classified by Suri (1984) as queueing network models, static allocation models, simulation models, perturbation analysis, and Petri-nets.

5.1.1. Queueing Network Models

Queueing network models are efficient tools for performance prediction of FMSs especially during early stages of design. Several models have been proposed to handle different features of an FMS. The basic one is due to Solberg, the earliest queueing network model of FMS, namely, CAN-Q (Solberg (1977)). CAN-Q is a queueing network model based on the assumptions of Baskett et al. (1975). It also assumes that blocking is not possible and machines are reliable. It is an evaluative model used as an aid to determine buffer size and machine groups.

In most of the cases, the output measures that queueing network models produce are average values which assume steady state operation of the system. However, the models tend to give reasonable estimates of performance, and are quite efficient; because they require relatively little input data, and do not use much computer time. Thus, queueing network models can be used interactively to quickly arrive at preliminary decision values. Most of the models that are mentioned in this section help in selecting the buffer size; however, analysis can be extended to deal with other design problems.

In FMS systems, it is desirable to provide some storage spaces to reduce the effects of breakdowns, variability in operation times, and blocking of machines due to the diversity of part routings. There are two basic alternatives which provide storage: local storage at machines or a common storage for the system which is accessible by all machines. Some systems use a combination of both local storage and common storage. To decide on the type of storage in an FMS, the system is represented as a network of queues.

For this purpose, Buzacott and Shanthikumar (1980) present a combination of restricted/unrestricted, open/closed queueing network models in which production capacity is chosen as the system performance measure. Given that transit times between machines and storage locations are negligible, these models show that common storage is superior to local storage. Local storage can be used, but only if there is a close control over the release of jobs to the system so that little blocking occurs. These models show the desirability of a balanced work load, i.e., the time that each part spends at each work station is same. The benefit of diversity in job routing is useful to increase the maximum capacity if there is adequate control on the release of jobs.

Yao (1983) considers an FMS with a set of stations linked by a recirculating conveyor. The system considered has the following characteristics :

1. Each station has a limited local storage,
2. Each station has a set of parallel machines with a general service-time distribution,
3. Jobs arrive at the system following a Poisson distribution,
4. Jobs are transported by the conveyor to the stations following a probabilistic routing,
5. Blocked jobs remain on the conveyor to be circulated. So, the recirculating conveyor also functions as a common storage, providing additional room for the stations,
6. Jobs finishing service at a station may either leave the system or be fed back to the conveyor according to a Bernoulli trial,
7. Whenever the congestion at the conveyor reaches a threshold, the external flow is turned off to ensure the occupancy by in-process jobs. The purpose of this policy is to control the congestion at the conveyor, to avoid "deadlock" and to give priority of occupancy to in-process jobs.

This system is modeled as an open restricted network of queues. In this model, machining times are taken to follow a general distribution, the randomness of which may be reflected by the squared coefficient of variation (SCV). The exponential distribution has a SCV of one. In FMS modeling, it is highly desirable that we can handle distributions with a large SCV. This need often arises in two occasions :

1. A wide variety of jobs are to be processed in the system,
2. Machine breakdowns are of such nature that they may rarely happen; the repair time will be much longer than the regular machining time of a job. If these factors are included into the aggregated service time distributions, its SCV is very likely to be greater than one.

Obviously, this open restricted network of queues with a general service time distribution cannot be modeled as a classical product-form network of queues, without losing most of the essential features of the FMS. So, an iterative scheme is developed to provide approximate solutions to the model. The focus of this model is on the storage aspect of the conveyor rather than the speed of the conveyor, because it is known from conveyor literature that the speed of the conveyor is not essential for the performance of the service stations.

The key to this model is the blocking probability of the external flow, b, which is equal to the probability that the number of jobs on the conveyor is the threshold value. If b is known, the total throughput of the system (TH) is immediately known and hence the throughputs of the conveyor (TH_0) and the stations (TH_i) are known, since a job will never be lost once it enters the system. Moreover, given the throughput of station i, TH_i, its behavior can be mimicked by an infinite waiting-room queue with an input rate TH_i (since no traffic is lost in an infinite waiting-room queue). From the queue length distributions of the set of mimic queues, the congestion at the conveyor can in turn be studied and a new b can be derived. It is not difficult to make this scheme iterative, so that starting from a rather arbitrary initial value of b, its actual value can eventually be found.

In summary, the above can be explained as follows. Define

$$\tilde{\lambda}_i = \text{adjusted input rate,}$$
$$TH_i = \text{throughput of station } i,$$
$$b_i = \text{fraction of the lost traffic in the limited queue,}$$

then,

$$\tilde{\lambda}_i = \frac{TH_i}{1 - b_i}$$

where b_i is iteratively solved as follows:

$$b_i^{(n+1)} = 1 - \frac{TH_i^{(n+1)}}{\tilde{\lambda}_i^{(n)}}$$

$$\tilde{\lambda}_i^{(n)} = \frac{TH_i}{1 - b_i^{(n)}}.$$

Once the adjusted input rates are found, classical results can be used to compute different performance measures. Convergence of the scheme is studied. A numerical example is also given, which compares results derived from the analytical model with those obtained through simulation. The performance measures used are throughput times, mean number of jobs in the system (and conveyor) and mean waiting time.

Yao and Buzacott (1986b) develop an approach that can be applied when stations in an FMS operate with non-exponential processing times. The idea is to transform the network of queues to an approximately equivalent exponential network, each station having a state-dependent service rate. The resultant network is an approximately equivalent open network which can be solved by the known techniques. Limited computational experience show that results are generally acceptable.

Bitran and Tirupati (1988) considered an unrestricted open queueing network with multiple product classes and deterministic routing. General distributions are assumed both for arrivals and service times. In case of a large number of products, the decomposition method (see the previous Section for this method) does not perform well to compute the performance measures such as mean number of jobs and queue lengths. Bitran and Tirupati consider the interference among products by aggregating all except the one in concern and represent it by an aggregate product. The two-product model, then, becomes the building block in the methodology. The squared coefficient of variations of product departures are used for aggregations. Three approximations are proposed for the computation of the interference effects. Each heuristic performs well for a different range of the squared coefficient of variation of the arrival distribution.

In another study performed by Suri and Diehl (1986), FMS is modeled as a closed restricted network which can undergo blocking. The performance measure is the throughput time. We know that existing closed queueing network analysis does not deal efficiently with the blocking problem in systems with finite buffers. On the other hand, detailed Markov models of systems with finite buffers become intractable for this problem. Yet, it is clear that the effect of blocking is significant in many real systems. This study addresses single class closed queueing networks with exponential service times and finite buffers, and uses approximation technique based on heuristic arguments. The approach involves modeling the rest of the network as "seen" by some server. To represent this view, the variable buffer size model is proposed. However, solving this model involves solving several two-server models. The method is compared with exact solutions or simulations, and found to be reasonably accurate. The method is easily implemented using standard software for closed queueing networks. Extension of the algorithm to give queue length distributions cannot be considered, because during the aggregation process the individual queue length information is lost.

In Yao and Buzacott (1986a), FMS is modeled as a closed restricted network. Specifically, modeling limited local buffers is focused on, because according to Yao and Buzacott, recent survey on FMSs indicate that most existing FMSs have a very small local buffer at the machines (usually one or two extra waiting spaces). This feature, however, has not been included in the classical closed network of queues model of FMSs which assumes unlimited local buffers at the stations. Special attention is given to the interpretation of the routing of parts and loading of work stations in these models, their relation to the operation of real system, and their impact on system performance. According to the way system is operated, one of the models can be applied. Although the models assume exponential processing times at the stations, their accuracies are good enough for all practical purposes, provided that the local buffers are substantially smaller than the total part population in the system. In the case of FMSs with large local buffers, the models can be used in combination with the exponentialization approach to yield accurate results.

In Yao and Buzacott (1985), FMS is modeled as a closed, restricted network, where parts routing follows a probabilistic shortest queue (PSQ) scheme, i.e., parts are routed to the shortest queue with the highest probability. The service times are exponential including the materials handling station which has travelling times, loading/unloading and refixturing operations. The central storage can accommodate N jobs where N is the total number of pallets available. It is proved that with the PSQ routing, the Markovian queue-length process satisfies time reversibility and has a product-form equilibrium distribution. The results obtained can be considered as the generalization of Towsley (1980). An algorithm is developed to compute the solution. The performance measures considered are throughput times, mean number of jobs and machine utilizations. This study exhibits a first effort to develop a model which accommodates some operational characteristics while still possessing a computationally tractable analytical structure.The last two queueing models discussed above can be used at the operational level of an FMS decision structure.

Yao and Buzacott (1987), in another study, consider a closed restricted queueing network. They derive closed-form solutions based on the theory of reversibility. An important case where reversibility holds is a system with a centralized material handling station.For this result to hold, the services are required to be exponentially distributed random variables. In the case where local buffers are zero, the product-form solutions are preserved for more general types of service distributions. For these reasons, the assumption of exponential service times is sufficient to obtain insights about a system where local buffer sizes are small.

Vinod and Altıok (1986) use an approximation for single item closed, unrestricted queueing networks based on two-stage representations. Service stations are subject to breakdown and repair. The approximation assumes exponentiality for the service completion times which has a two-stage coxian distribution. The two-stage representation is exact, but extension of the idea to multiple stages creates the approximations. Numerical results indicate that the approximate throughput and mean queue length figures are sufficiently close to their exact values.

Queueing network models, as a decision aid for FMS designers and managers, are mostly appropriate for planning and initial design phases as discussed above. They are evaluative models rather than being generative.

5.1.2. Models Based on MVA

Suri and Hildebrant (1984) developed Mean Value Analysis of Queues, MVAQ. MVAQ is a computer program based on the mean value analysis of queues. MVAQ's contribution to FMS modeling is to extend the basic functionality of CAN-Q with some additional features; possibility of modeling multiple part classes, low computer memory requirements, and more accurate computation of the performance of large systems.

An important point to be addressed is that the theory behind MVAQ is based on several assumptions which are usually not satisfied by an FMS. Examples of these are: (a) the processing time of a part has an exponential probability distribution (in an FMS, the processing times are usually known quite accurately), and (b) the routing of a part to the next machine is chosen probabilistically (in an FMS, the next operation/machine is often predetermined). Nevertheless, quite reasonable predictions of FMS performance can be obtained using MVA techniques. This has been explained to some extent by Denning and Buzen (1978) with a new theoretical approach called operational analysis. A rigorous analysis of the alternative assumptions of operational analysis has recently been done by Suri (1983) which justifies the use of MVA techniques for practical systems such as an FMS. Based on these new theoretical developments, as well as extensive practical usage, it can be said that MVA provides predictions of FMS performance that are reasonable for planning and control decisions.

There are certain conditions under which MVAQ does not work very well. This is when there is considerable congestion in the material transport and/or material buffer systems, leading to a significant amount of blocking. If blocking accounts for a significant proportion of the total waiting time at machines in a system, MVAQ may seriously overestimate the system's production rate.

MVAQ models the behavior of an FMS in an undetailed fashion. The method used (queueing theory) aggregates and approximates the occurrence of the activities such as the movement of particular parts, and the machining of particular operations. Thus, MVAQ should be used to arrive at "first-cut" decisions on both the design and operation of an FMS. Determining the optimum number of machines for an FMS, the minimum number of pallets/fixtures needed for production, the best routing for parts, and many other issues, can be approached using MVAQ. Whenever more accurate results are needed, some other models, i.e., simulation, Perturbation Analysis, should be employed to "fine tune" the detailed decisions.

Seidmann et al. (1986) compare five evaluative models for FMSs on a qualitative basis. The methods compared are CAN-Q by Solberg (1977), MVA by Reiser and Lavenberg (1980), and two extensions of MVA with a simulation model. Advantages and restrictions of all models are listed. In a complementary study, Shalev-Oren et al. (1985) numerically evaluate the above models to indicate their accuracies. It turns out that in the system configurations evaluated, there are no significant differences among different approaches in terms of throughput, utilizations, queue length and waiting time estimates. However, the deviation in estimating throughput and utilizations are smaller than queue length and waiting time estimates, indicating that queueing network models are more suitable for design problems, rather than for operational ones.

5.1.3. Some Other Models

At present, simulation is perhaps the most widely used computer-based performance evaluation tool for FMS, although it is more expensive than other evaluation methods for FMS. There is the possibility to use powerful simulation languages or even some simulation packages which are specially tailored to FMS to ease the model-building and data input effort (see Akella et al.(1985), Foley and Jain (1987), Montazeri et al. (1988), Jain and Foley (1986) and Suri (1984). An example of how simulation is used in an FMS environment is given by Co et al. (1988). A number of sequencing rules are evaluated to see their effects on the queue-lengths of work stations. Hybrid Simulation/Analytical models are also encountered in the literature (see Shanthikumar and Sargent (1980)).

Another method, Perturbation Analysis (P/A) to discrete event systems, can be applied to FMSs. The basic idea is to observe the detailed behavior of the system, whether through simulation or from the actual system, for one set of decision parameters. By doing some minor additional calculations while the system is being observed, P/A can predict the system behavior if these decisions were changed. The important point is that it is not necessary to re-run the system (or simulation), that is, all predictions are obtained from one observation. Thus, observation of one experiment can give accurate directions for the improvement of several parameter values. This provides an efficient method for on-line optimization of complex systems. The main disadvantage of P/A is that it currently cannot predict accurately the effects of "large" changes in decisions (see Suri (1984), Suri and Dille (1984)).

Petri Net is another tool used for the analysis of FMSs. For the specification of FMSs as Petri Nets see Martinez et al. (1986). An example modeling and analysis of FMSs using a Petri Net approach is given by Narahari and Viswanadham (1985).

As a result, we can conclude that evaluative models discussed above are more of a tool to help the decision maker sharpen his/her intuition about the system; they provide insight rather than decisions. It may require considerable effort to find good solutions using such models. So, in this respect, evaluation models are semi-generative models. Especially, the perturbation analysis technique could well be considered an evaluative and semi-generative model.

5.2. Generative Models for FMSs

An important area of study in queueing network formulations of FMSs is to show a kind of monotonicity of queue lengths with respect to certain system parameters. Suri (1985), and Shanthikumar and Yao (1987) are two such examples. The results obtained are very important in constructing generative techniques. However, these studies still lack the details needed by a queueing network model of FMS. In this subsection, we discuss a number of models which we find useful in reviewing generative models of queueing networks for FMSs.

5.2.1. Vinod and Solberg (1985) and Dallery and Frein (1986)

A frequently encountered design issue for an FMS is to find the lowest cost configuration, i.e., numbers of resources of each type (machines, pallets, etc.) which yield a given production rate. A cost is associated with each resource type and the best configuration is the one with the lowest overall cost. This can be formulated as an optimization problem. This problem has been studied in a recent paper by Vinod and Solberg (1985).

Vinod and Solberg (1985) model FMS as a closed network of queues, for which the optimal cost effective system configuration is determined by a partial implicit enumeration algorithm.

The problem formulation to find the optimal system configuration to achieve the desired system throughput may be mathematically stated as follows:

$$\text{Minimize} \quad Z = \sum_{i=1}^{M} k_i c_i + k_N N$$
$$\text{s.t.} \quad X \geq XP$$
$$c_i \; \forall i, N \text{ integer}$$

where k_i, $(i = 1, 2, , ..., M, N)$ is the cost per unit including operating costs and capital investment of a server or a job, X is the actual system throughput for a given system configuration, XP is the desired system throughput, and N is the number of customers allowed in the system.

Solving the above integer program is formidable since the nonlinear constraint cannot be explicitly represented in terms of the decision variables. Precise evaluation of the actual system throughput, X ,via closed queueing network analysis is only possible when the values for the decision variables that define the system configuration are specified. Since the performance measures are in closed form only in a very abstract sense, optimization in closed queueing networks is very difficult if not impossible. Vinod and Solberg conducted an experimental investigation of optimizing parameters in closed unrestricted queueing networks by using direct search techniques.

This methodology to design the optimal system configuration of FMSs is studied for both reliable and unreliable systems. For the reliable system, the constraint in the integer program above is evaluated by performing a regular product-form analysis. But, when the resources are unreliable, the steady state probabilities of closed network are no longer of product-form. Assuming that service, failure and repair times are all exponentially distributed, the service completion time for jobs at each resource has a two-stage Coxian distribution (see Altıok and Stidham (1983)). Evaluation of the constraint is accomplished by the exponential approximation (EXP) wherein exponentiality is assumed for the service completion time of two-stage Coxian distribution. Detailed experimental validation suggests that EXP overestimates the actual system throughput to within an accuracy of 5% for a wide range of parameters.

So, this methodology may be used to aid accurately in planning and design phases of FMSs which are closed, unrestricted and reliable/unreliable.

In Dallery and Frein (1986), an efficient method to determine the optimal configuration is presented. It is an alternative procedure to Vinod and Solberg (1985) to obtain the optimal configuration. The FMS is modeled by a closed single class queueing network, consisting of M stations and N customers. At station i, there are C_i servers, $i = 1, ..., M-1$. Station M models the material handling system(MHS) and C_M is the number of carts available. The number of customers N, is equal to the number of pallets available in the system. The characteristic parameters of the closed queueing network are as follows :

S_i = mean service time at station i,
V_i = visit ratio at station i (average number of times a customer visits station i between two successive visits at the loading/unloading (L/U) station).

Then, $Y_i \ (= V_i S_i)$ is called the loading of station i.

To find the optimal system configuration, the problem is stated mathematically as follows :

Minimize $Z = (\underline{X})$
s.t. $X \geq XP$

where $\underline{X}$ integer decision variable vector, and

XP = prescribed system throughput (desired production rate of the FMS),

X = actual system throughput of the network for a given system configuration,

$$\underline{X} = \begin{cases} (C_1, ..., C_M) & , \text{if } N \text{ is fixed} \\ (C_1, ..., C_M, N) & , \text{if } N \text{ is a decision variable} \end{cases}$$

Z = actual configuration cost (increasing function),

The solution procedure is composed of three steps :

1. Determine a lower configuration

The goal is to find a lower configuration which is close to the optimal configuration. For this purpose they use some well known properties of closed queueing networks provided by bottleneck analysis (Denning and Buzen (1978)) (so-called asymptotic bound analysis (ABA) in the paper). This analysis provides upper bounds on the throughput of general closed queueing networks. These bounds are given by :

$$X \leq \min \frac{C_i}{Y_i}$$

and

$$X \leq \frac{N}{R_0}$$

where R_0 is the total loading defined by $R_0 = \sum_{i=1}^{M} Y_i$.

These upper bounds are obtained considering the extremes of system behaviour: either no queueing delay or at least one station is saturated. From these bounds and prescribed throughput, the following lower bounds on the decision variables can be obtained :

$$C_i > XP \cdot Y_i$$

for any i, and

$$N \geq XP \cdot R_0 \ .$$

The solutions $(C_1, ..., C_M$ and $N)$ which are greater than the lower bounds give lower configurations. Such lower configurations are considered during the optimal search procedure, as any configuration with $C_i \leq XP \cdot Y_i$ for any i and/or $N < XP \cdot R_0$ is not going to give the prescribed throughput.

2. Heuristic Solution

In order to reduce the number of throughput evaluations required during the optimal solution search procedure, and speed up the optimization, a "good" starting solution is needed. The initial configuration is provided by the lower server configuration derived in step i. Then at each step of this procedure , the number of servers at one station is increased by one. The station is the one which maximizes the ratio of the extra throughput to the extra cost, due to the additional server. The procedure is stopped as soon as the prescribed system throughput is achieved. In case the number of customers is a decision variable, then, at each step, either a server or a customer would be added.

3. Optimal Solution

In order to get the optimal solution, an implicit enumeration algorithm is used. The implicit enumeration algorithm starts with the lower configuration and then increases the decision variables to systematically enumerate all possible configurations.

5.2.2. Stecke and Solberg (1985), Stecke and Morin (1985), and Shanthikumar and Stecke (1986)

Stecke and Solberg (1985) considered FMS as a closed network of queues. The main assumptions are:

1. There are always n parts in the system,
2. Each machine group contains sufficient room for n parts, so that there will be no blocking or starving of machines,
3. The part follows a probabilistic routing.

Answers for the following two questions are investigated:

1. Given the freedom to arrange the available servers into groups of varying size, is it optimal to equalize the number of servers in each group ?
2. Given a specific server configuration (equal or unequal), what is the optimal workload/server?

The results are interesting since they can be considered as counter-intuitive:

1. Given that each group has the same number of machines, the expected production rate is maximized when the allocation per machine (i.e., workload) is balanced,
2. Given a system of groups of pooled machines of unequal sizes, the production rate is maximized by a unique allocation which is unbalanced. Specifically, more (less) than the balanced amount of work per machine is assigned to the larger (smaller) groups of machines,
3. If the number of parts allowed, n, is infinite, then for **any** sized machine groups, optimal allocation is balanced.

As a result, machine groupings can be ordered according to the maximum expected production rate if the assumptions of the model hold. The numerical results tabulated indicate that as the number of machines available increases, the potential of benefits in applying the above results increases.

Stecke and Morin (1985) study an idealized version of the FMS loading problem, which is one of the set-up problems of an FMS. In the context of a closed queueing network, the loading problem involves determining the best allocation of operations and associated cutting tools of a set of part types among the machine tools subject to technological and capacity constraints, so as to maximize expected production. In a previous paper (Stecke and Solberg (1985)), as discussed above, it was shown that unbalancing the number of machines in each machine group is the best way to partition the machine tools of a particular type into machine groups; and for these unbalanced configurations, expected production is maximized by a particular unbalanced allocation of workload per machine. In this paper, however, it is shown that balancing the workload per machine can be best even in some systems with grouped machines because of the discreteness of operation times, different machine tool requirements, and limited capacity tool magazines. This paper characterizes situations in which balancing is optimal: for those systems in which there is no grouping or pooling of machines of similar type into machine groups. A single-server closed queueing network model is used to analyze the optimality of balancing for adequately buffered FMSs (n parts circulating, and a buffer capacity to hold n parts at each machine tool) in which each operation is assigned to only one machine. It is shown that balancing maximizes the expected production in these systems. Specifically, symmetric mathematical programming and generalized concavity is used to establish the optimality of balanced workloads.

The results obtained in Shanthikumar and Stecke (1986) complement those obtained in Stecke and Morin (1985). The former one establishes that maintaining a balanced workload on each machine over time stochastically minimizes the work-in-process inventory in certain types of FMSs while the latter establishes that balancing workloads maximizes expected production again for the same particular types of FMSs. However, the scenario in this paper is different than it is in the previous similar studies, Stecke and Morin (1985) in particular. An open queueing network is used to model the random arrival of individual parts to the FMS. Given that there are N parts in the system, a closed queueing network then provides the expected production, if the workload on each machine is also provided. Three strategies are considered to release the parts, which have arrived into a production control area, into the FMS. Under the three release policies, if the workload per machine remains balanced over time, not only is the expected production maximized, but the in-process inventory is stochastically minimized. There is no pooling of machines and there is a finite or infinite common buffer area for incoming parts (Poisson arrival) which are subsequently released based on the three dispatch policies. So, the problem is dynamic unlike the previous studies in which the problem is static.

5.2.3.　Wang (1986)

The FMS considered is modeled as on open restricted network where standard product-form of the steady state probabilities can be obtained. The machines are assumed to have equal load. It is also assumed that an (s, Q) policy can be implemented where s is the reorder point defined as the number of parts in the system and Q is the additional number of parts brought instantaneously into the system as a batch. The system has M machines and N pallets and it is assumed that no blocking or starving will occur. Then the number of parts in the system at any time is in between s and s+Q. The throughput rate of this system is a function of the number of parts in the system and is given by Zahorjan et al. (1982). It is a simple formula and given various costs (transportation cost per trip from the central storage to the flexible cell, fixed cost of the local buffer at the cell per unit time and unit carrying cost per unit time) and profit per part, a model can be built. The solution is hard as it requires a search procedure. To facilitate the solution procedure, a power approximation method is used. Power approximation method is to represent s and Q as regression equations which are based on the exact solution of a wide variety of models. The results obtained are only valid within the range of the parameters used in problems solved for power approximation. The quality of fit is a function of the number of exact solutions obtained (in terms of the regression analysis, the number of exact solutions is the number of data points).

5.2.4.　Vinod and Sabbagh (1986)

The model presented is a hybrid model as it contains two different stages :

1. A performance model which measures the performance of an FMS when tool failures and repair times are considered,
2. An optimization model which minimizes cost of the repairing facility subject to a certain availability level. The allocation of spares will indicate a certain tool configuration that will have a certain performance which can be measured by the performance model.

The FMS considered is of simple type which does not allow for blocking and starvation as the system is considered as a closed unrestricted queueing network with multiple classes. The regular product-form analysis would not work as the total availability factor is restricting. If there are no spare tools available at a given station, the mean service time of the station can be considered the mean time to tool failure. By keeping spare tools, the mean service time of the station can be decreased. Assuming that the steady state probability distribution for tool availability for a station dictates the steady state processing rate of the station, load dependent transformation can be used. Load dependent transformation is the representation of the service time of a station which includes the extended time caused by the absence of the tool. The approach is validated by some experiments.

To determine the spare levels an optimization model based on the following two assumptions are used :

1. The number of channels in the repair facility is infinite (ample service assumption),
2. The Poisson failure process of the tools are not state dependent and the calling population is infinite.

These assumptions will result in statistically independent resupply times for the tools. The objective is to minimize the sum of cost of spares and cost of capacity of the repair facility. The minimization is subject to a constraint on the marginal queue length distribution of the tool.

The solution methodology is based on the fact that the constraints are monotonically nondecreasing in the decision variables.

The hybrid approach comes into picture where we take the solution of the optimization problem and measure the performance using the performance model. We repeat the procedure by solving problems for different queue length distributions of the tools. No information is given to clarify this step, so we cannot comment on the effectiveness of the procedure.

5.2.5. Sabbagh and Vinod (1984)

The system considered assumes a closed unrestricted queueing network. Each station has a single server with a service time distribution that is exponentially distributed. The transporter is defined as one of the stations and the state of the system is defined by the number of workparts at each station. The performance measure that is considered is the throughput of the system. The objective is to determine the optimal part-mix ratio in the closed network which minimizes the total cost subject to a lower bound on the system throughput. The objective function can be linear or nonlinear, but should be isotone nondecreasing in all decision variables (isotone nondecreasing functions are monotone nondecreasing functions that require monotonicity on each variable).

An implicit enumeration technique is used to solve the problem. Even for small sized problems the processing time is quite large and the number of experiments solved is not sufficient to give an idea on the efficiency of the technique.

5.2.6. Yao and Shanthikumar (1986)

Yao and Shanthikumar (1986) consider an FMS system that consists of a set of cells where each cell processes a certain type of part family. In this paper, the optimal allocation of buffer space, machines, and the production capacity of the upstream production stage is studied. A marginal allocation approach is used to solve the problem of optimal allocation of buffers and machines efficiently. The problem of optimal allocation of production capacity turns out to be a problem to minimize a convex function over a convex set.

5.2.7. Kayalıgil (1989), and Avonts and Van Wassenhove (1988)

Kayalıgil (1989) handles part-mix allocation problem in FMSs. The overall part-mix of a production system is optimally allocated between the existing conventional facilities and planned automated system to be acquired. So, total cost of production is broken up into two parts: the cost for producing parts in the automated facilities and the cost for complementing production in the conventional shop. For a given part-mix, two main sources of costs are identified as cost of work-in-process inventories carried during manufacturing and cost of time spent in manufacturing. Other sources of costs are either neglected or assumed to be indifferent to part allocation between the two facilities. Mean value analysis (MVA) is utilized to evaluate performance of the automated system under trial part-mix allocations; mean values for the throughput and queue lengths are thus obtained. These two mean values depend on the number of each part type allocated to a route, and these allocations in turn affect manufacturing costs. Given the postulated part-mix for automated manufacturing, optimal allocation to routes is performed by a branch and bound procedure.

Avonts and Van Wassenhove (1988) handle part-mix and routing mix problem in FMSs. They consider an FMS to have mainly two flexibilities: part mix flexibility, i.e., which part types should be produced on FMS and in what quantities, and routing mix flexibility, i.e., possibility of assigning operations to different machines. The problem is stated as determining the number of parts of each type to be produced on the FMS in period T and the machines on which the operations should be performed, such that cost savings (compared to production of these parts on the conventional system) is maximized subject to capacity and demand constraints.

Their proposed solution consists of an hierarchical combination of an LP model and a queueing network model. The LP model determines a static solution (which part types to produce on the FMS and in what quantities together with the assignment of operations to machines) based on operation times and available machine capacity. This solution is then examined with a queueing model to evaluate some dynamic effects (waiting lines, transport system). This information is then fed back to the LP model in order to obtain a more realistic solution. In this paper, they use CAN-Q model. The input consists of different routing descriptions, number of parts in the system, number of transporters, average transportation time and the mix of each part type; and the output contains a lot of aggregate information about the behaviour of the system such as production rate, average waiting times, average utilization of the machines.

6. CONCLUSIONS

An FMS is such a complex system that it becomes very hard to fully describe its behavior via a single analytical model that we can solve. Thus, simulation has been the dominant tool for planning, design, and operation of FMSs for long. While simulation models are of great value for evaluative purposes, analytical models seem to be superior in terms of the insight that they give.

Queueing networks have become more important for performance evaluation of complex systems such as FMSs. Baskett et al. (1975) have shown that the product-form solutions exist for open, closed, and mixed queueing networks with infinite buffer spaces, different job classes, nonexponential service time distribution, and different queueing disciplines. But still, the queueing network models are not adequate in planning of a real FMS which has a finite buffer space, both local and common, complex scheduling and routing schemes, and machine breakdowns. Current queueing network analyses do not deal efficiently with the problem of blocking in queueing systems with finite buffers. As stated above, the product-form networks do not accurately account for the effect of finite buffers. On the other hand, detailed Markov models for this class of systems become intractable. For this reason, several approximate methods have been proposed to deal with the finite buffer problem in queueing networks. However, to better represent the operation of FMS, the existing analytic and approximate models need to be extended in a number of directions.

Another research area is the optimization of the closed networks that was already accomplished in an implicit enumeration framework by exploiting the monotonicity of the closed network. This also seems to be an interesting area for further research.

A new research area in planning, and design issues of FMSs has been towards improvement of hybrid simulation/analytic models which seem to give quite good results and are cost effective. These hybrid models aim at modeling such FMSs for which analytic or acceptable approximate results are not available.

7. REFERENCES

1. Akella, R., Bevans, J.P., and Choong,Y. 1985. Simulation of a Flexible Manufacturing System. Laboratory for Information and Decision Systems, LIDS-P-1435, MIT, Cambridge, Massachusetts.

2. Altıok, T. 1985. Approximate Analysis of Production Lines with General Service and Repair Times and with Finite Buffers. Rutgers University, Industrial Engineering Department, Piscataway, New Jersey.

3. Altıok, T., and Stidham, S. 1983. The Allocation of Interstage Buffer Capacities in Production Lines. *IIE Trans.* **15**, 292-299.

4. Avonts, L.H., and Van Wassenhove, L.N. 1988. The Part Mix and Routing Mix Problem in FMS: A Coupling Between an LP Model and a Closed Queueing Network. *Inter. J. Prod. Res.* **26**, 1891-1902.

5. Baskett, F., Chandy, K.M., Muntz, R.R., and Palacios, F.G. 1975. Open, Closed, and Mixed Networks of Queues with Different Classes of Customers. *JACM* **22**, 248-260.

6. Bessant, J. 1984. Flexible Manufacturing Systems: an Overview. UNIDO, Microelectronics Monitor, No.12 Supplement, UNIDO.

7. Bitran, G.R., and Tirupati, D. 1988. Multiproduct Queueing Networks with Deterministic Routing: Decomposition Approach and The Notion of Interference. *Mgmt. Sci.* **34**, 75-100.

8. Black, J.T. 1983. Cellular Manufacturing Systems Reduce Setup Time, Make Small Lot Production Economical. *Ind. Eng.* **15**, 36-48.

9. Brandwajn, A., and Jow, Y.L. 1988. An Approximation Method for Tandem Queues with Blocking. *Opns. Res.* **36**, 73-83.

10. Browne, J., Dubois, D., Rathmill, K., Sethi, S.P., and Stecke, K.E. 1984. Classification of Flexible Manufacturing Systems. *The FMS Magazine* **2**, 114-117.

11. Buzacott, J.A. 1967. Automatic Transfer Lines with Buffer Stocks. *Inter. J. Prod. Res.* **5**, 183-200.

12. Buzacott, J.A., Shanthikumar, J.G. 1980. Models for Understanding Flexible Manufacturing Systems. *AIIE Trans.* **12**, 339-350.

13. Buzacott, J.A., and Yao, D.D.W. 1986. Flexible Manfacturing Systems: Review of Analytical Models. *Mgmt. Sci.* **32**, 890-905.

14. Buzen, J.P. 1973. Computational Algorithms for Closed Queueing Networks with Exponential Servers. *CACM* **16**, 527-531.

15. Co, H.C., Jaw, T.J., and Chen, S.K. 1988. Sequencing in Flexible Manufacturing Systems and Other Short Queue-Length Systems. *J. Manuf. Sys.* **7**, 1-9.

16. Co, H.C., and Wysk, R.A. 1986. The Robustness of CAN-Q in Modeling Automated Manufacturing Systems. *Inter. J. Prod. Res.* **24**, 1485-1503.

17. Cook, N.H. 1975. Computer-Managed Parts Manufacture. *Scientific American* **232**, 28-38.

18. Curtin, F.T. 1983. Automating Existing Facilities: GE Modernizes Dishwasher, Transportation Equipment Plants. *Ind. Eng.* **15**, 32-38.

19. Dallery, Y., and Frein, Y. 1986. An Efficient Method to Determine the Optimal Configuration of a Flexible Manufacturing System. *Proceedings of the Second ORSA/TIMS Conference on Flexible Manufacturing Systems* , 269-282.

20. Denning, P.J., and Buzen, J.P. 1978. The Operational Analysis of Queueing Network Models. *Computing Surveys* **10**, 225-261.

21. Foley, W.J., and Jain, S. 1987. Using Simulation Generators for Modeling Flexible Manufacturing Systems. Technical Report No. 37-87-137, Rensselaer Polytechnic Institute, New York.

22. Erke, Y., and Erkip, N. 1986. Overview of Several Approaches to Hierarchical Planning in FMS. Technical Report No. 86-04, Department of Industrial Engineering, METU, Ankara,Turkey.

23. Gershwin, S.B. 1987A. An Efficient Decomposition Method for the Approximate Evaluation of Tandem Queues with Finite Storage Space and Blocking. *Opns. Res.* **35**, 291-305.

24. Gershwin, S.B. 1987B. Representation and Analysis of Transfer Lines with Machines That Have Different Processing Rates. *Ann. Opns. Res.* **9**, 511-530.

25. Gershwin, S.B., and Schick, T.C. 1983. Modeling and Analysis of Three-Stage Transfer Lines with Unreliable Machines and Finite Buffers. *Opns. Res.* **31**, 354-380.

26. Gordon, W.J., and Newell, G.F. 1967. Closed Queueing Systems with Exponential Servers. *Opns. Res.* **15**, 254-265.

27. Groover, M.P. 1980. Automation, Production Systems, and Computer- Aided Manufacturing. *Flexible Manufacturing Systems*. Prentice Hall, Englewood-Cliffs, New Jersey.

28. Heywood,P. 1983. Demo FMS Produces Round Parts. *American Machinist* **127**, 100-102.

29. Ho,Y.C., Eyler, M.A., and Chien, T,T 1979. A Gradient Technique for General Buffer Storage Design in a Production Line. *Inter. J. Prod. Res.* **17**, 557-580.

30. Jablonowski, J. 1985. Keeping an FMS up to Date. *American Machinist* **129**, 76-78.

31. Jackson, J.R. 1957. Networks of Waiting Lines. *Oper. Res.* **5**, 518-521.

32. Jackson, J.R. 1963. Jobshop-Like Queueing Systems. *Mgmt. Sci.* **10**, 131-142.

33. Jain, S., and Foley, W.J. 1986. Basis for Development of A Generic FMS Simulator. *Proceedings of the Second ORSA/TIMS Conference on Flexible Manufacturing Systems*, 393-403.

34. Kayalıgil, M.S. 1989. Part-Mix Allocation Between Automated and Conventional Manufacturing. Technical Report No: 89-10, Department of Industrial Engineering, Middle East Technical University, Ankara, Turkey.

35. Koenigsberg, E. 1982. Twenty Five Years of Cyclic Queues and Closed Queue Networks: A Review. *J. Opnl. Res. Soc.* **33**, 605-619.

36. Kusiak, A. 1986. Application of Operational Research Models and Techniques in Flexible Manufacturing Systems. *Eur. J. Opnl. Res.* **24**, 336-345.

37. Martinez, J., Alla, H., and Silva, M. 1986. Petri Nets for the Specification of FMSs. In *Modeling and Design of Flexible Manufacturing Systems* , A.Kusiak (Ed.), North Holland, Amsterdam, 389-406.

38. Montazeri, M., Gelders, L.F., and Van Wassenhove, L.N. 1988,. A Modular Simulator for Design, Planning, and Control of Flexible Manufacturing Systems. *Inter. J. Adv. Manuf. Tech.* **3**, 15-32.

39. Narahari, Y., and Viswanadham, N. 1985. A Petri Net Approach to the Modeling and Analysis of Flexible Manufacturing Systems. *Ann. Opns. Res.* **3**, 449-472.

40. Perros, H.G.,Nilsson, A.A., and Liu, Y.C. 1985. Approximate Analysis of Product-Form Type Queueing Networks with Blocking and Deadlock. Computer Science Report, North Carolina State University, Releigh, North Carolina.

41. Reiser, M.,and Lavenberg, S.S. 1980. Mean-Value Analysis of Closed Multichain Queueing Networks. *JACM* **27**, 313-323.

42. Sabbagh, M.,and Vinod, B. 1984. On Optimal Product Mix Ratios to Maximize Productivity. SUNY College of Technology, Department of Computer Science, New York.

43. Schweitzer, P.J.   1977. Maximum Throughput in Finite-Capacity Open Queueing Networks with Product-Form Solutions. *Mgmt. Sci.* **24**, 217-223.

44. Seidmann, A., Shalev-Oren, S., and Schweitzer, P.J. 1986. An Analytical Review of Several Computerized Closed Queueing Network Models of FMS. *Proceedings of the Second ORSA/TIMS Conference on Flexible Manufacturing Systems.*

45. Shalev-Oren, S., Seidmann, A., and Schweitzer, P. 1985. Computerized Closed Queueing Network Models of Flexible Manufacturing Systems: A Comparative Evaluation. Working Paper Series No. QM 8548, The Graduate School of Management, University of Rochester, Rochester, New York.

46. Shanthikumar, J.G., and Sargent, R.G. 1980. A Hybrid Simulation Analytic Model of a Computerized Manufacturing System. Working Paper No.80-017, Syracuse University, Department of Industrial Engineering and Operations Research, Syracuse, New York.

47. Shanthikumar, J.G., and Stecke, K.E. 1986. Reducing Work-In-Process Inventory In Certain Classes of Flexible Manufacturing Systems. *Eur. J. Oprl. Res* **26**, 266-271.

48. Shanthikumar, J.G., and Yao, D.D. 1987. Stochastic Monotonicity of the Queue Lengths in Closed Queueing Networks. *Opns. Res.* **35**, 583-588.

49. Solberg, J.J. 1977. A Mathematical Model of Computerized Manufacturing Systems. *Proceedings of the Fourth International Conference on Production Research*, Tokyo, Japan.

50. Stecke, K.E. 1986. A Hierarchical Approach to Solving Machine Grouping and Loading Problems of Flexible Manufacturing Systems. *Eur. J. Oprl. Res* **24**, 369-378.

51. Stecke, K.E. 1988. O.R. Applications to Flexible Manufacturing. In *Operational Research '87* , G.K.Rand (Ed.), Elsevier, Amsterdam, 217-232.

52. Stecke, K.E., and Morin, T.L. 1985. The Optimality of Balancing Workloads in Certain Types of Flexible Manufacturing Systems. *Eur. J. Oprl. Res* **20**, 68-82.

53. Stecke, K.E., and Solberg, J.J. 1985. The Optimality of Unbalancing Both Workloads and Machine Group Sizes in Closed Queueing Networks of Multiserver Queues. *Opns. Res.* **33**, 882-910.

54. Suri, R. 1983. Robustness of Queueing Network Formulas. *JACM* **30**, 564-594.

55. Suri, R. 1984. An Overview of Evaluative Models for Flexible Manufacturing Systems. *Proceedings of First ORSA/TIMS Conference on Flexible Manufacturing Systems.*

56. Suri, R. 1985. A Concept of Monotonicity and its Characterization for Closed Queueing Networks. *Opns. Res.* **33**, 606-624.

57. Suri, R., and Diehl, G.W. 1986. A Variable Buffer-Size Model and Its Use in Analyzing Closed Queueing Networks with Blocking. *Mgmt. Sci.* **32**, 206-224.

58. Suri, R., and Dille, J.W. 1984. On-Line Optimization of Flexible Manufacturing Systems Using Perturbation Analysis. *Proceedings of First ORSA/TIMS Conference on Flexible Manufacturing Systems.*

59. Suri, R., and Hildebrant, R.R. 1984. Modeling Flexible Manufacturing Systems Using Mean-Value Analysis. *SME J. Manuf. Sys.* **3**, 27-38.

60. Suri, R., and Whitney, C.K. 1984. Decision Support Requirements in Flexible Manufacturing. *J. Manuf. Sys.* **3**, 61-69.

61. Takahashi, Y., Miyahara, H., and Hasegawa, T. 1980. An Approximation Method for Open Restricted Queueing Networks. *Opns. Res.* **28**, 594-602.

62. Towsley, D. 1980. Queueing Network Models with State-Dependent Routing. *JACM* **27**, 323-337.

63. Vinod, B., and Altıok, T. 1986. Approximating Unreliable Queueing Networks Under the Assumption of Exponentiality. *J. Opnl. Res. Soc.* **37**, 309-316.

64. Vinod, B., and Solberg, J.J. 1985. The Optimal Design of Flexible Manufacturing Systems. *Inter. J. Prod. Res.* **23**, 1141-1151.

65. Vinod, B., and Sabbagh, M. 1986. Optimal Performance Analysis of Manufacturing Systems Subject to Tool Availability. *Eur. J. Opnl. Res.* **24**, 398-409.

66. Von Looveren, A.J., Gelders, L.F., and Van Wassenhove, L.N. 1986. A Review of FMS Planning Models. In *Modeling and Design of Flexible Manufacturing Systems*, A.Kusiak (Ed.), Elsevier, Amsterdam. 3-31.

67. Wang, H. 1986. Analysis of an FMS Design Problem by Power Approximation. In *Modeling and Design of Flexible Manufacturing Systems*, A.Kusiak (Ed.), Elsevier, Amsterdam, 301-308.

68. Yao, D.D. 1983. Modeling a Flexible Manufacturing System with a Recirculating Conveyor and Limited Local Storages. University of Toronto, Department of Industrial Engineering, Toronto.

69. Yao, D.D., and Buzacott, J.A. 1985. Modeling a Class of State-Dependent Routing in Flexible Manufacturing Systems. *Ann. Opns. Res.* **3**, 153-167.

70. Yao, D.D., and Buzacott, J.A. 1986a. Models of Flexible Manufacturing Systems with Limited Local Buffers. *Inter. J. Prod. Res.* **24**, 107-118.

71. Yao, D.D., and Buzacott, J.A. 1986b. The Exponentialization Approach to Flexible Manufacturing System Models with General Processing Times. *Eur. J. Opnl Res* **24**, 410-416.

72. Yao, D.D., and Buzacott, J.A. 1987. Modeling a Class of Flexible Manufacturing Systems with Reversible Routing. *Opns. Res.* **35**, 87-93.

73. Yao, D.D., and Shanthikumar, J.G. 1986. Some Resource Allocation Problems In Multi-Cell Systems. *Proceedings of the Second ORSA/TIMS Conference on Flexible Manufacturing Systems.*

74. Zahorjan, J., Eager, D.L., and Sweillam, H.M. 1988. Accuracy, Speed 4, and Convergence of Approximate Mean Value Analysis. *Performance Evaluation* **8**, 225-270.

75. Zahorjan,J., Kenneth, C.S., Eager, D.L., and Galler, B. 1982. Balanced Job Bound Analysis of Queueing Networks. *CACM* **25**, 134-141.

Index